U.S. MARINES Close-Quarters Combat Manual

U.S. MARINES Close-Quarters Combat Manual

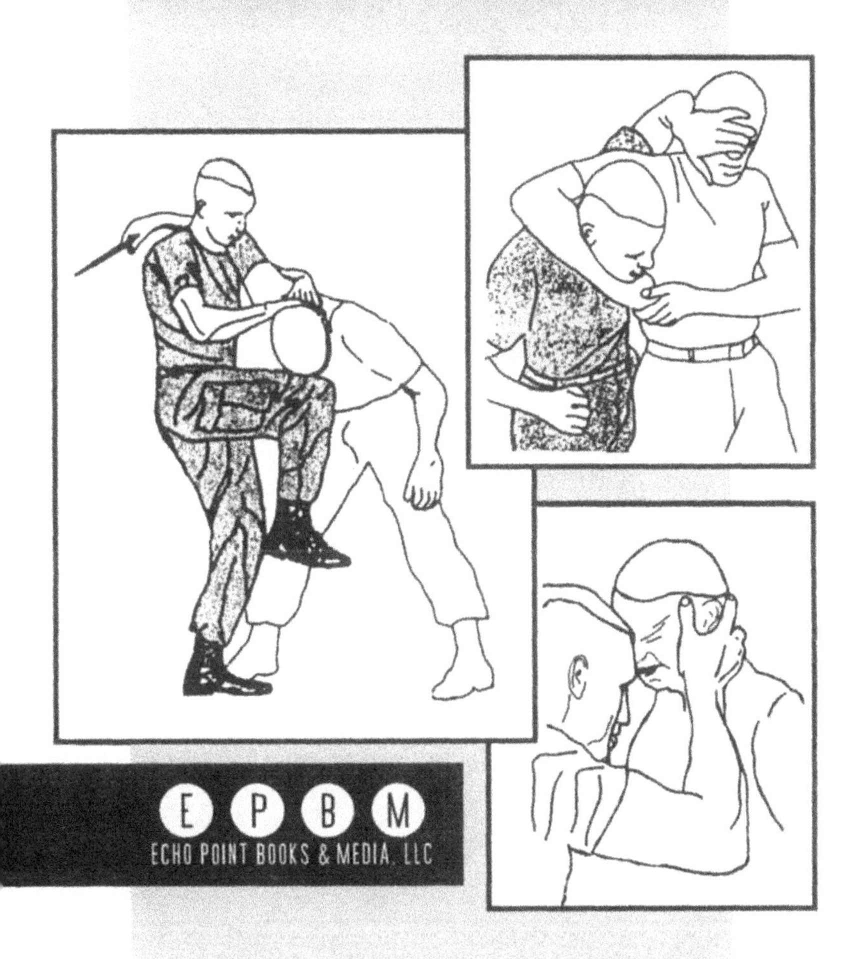

Published 2016 by Echo Point Books & Media
Brattleboro, Vermont
www.EchoPointBooks.com

U.S. Marines Close-Quarters Combat Manual
ISBN: 978-1-62654-499-4 (paperback)
978-1-62654-514-4 (casebound)
978-1-62654-515-1 (spiralbound)

Cover image courtesy of U.S. Marine Corps
Photo by Cpl. Joshua Hines

Editorial and proofreading assistance by Ian Straus,
Echo Point Books & Media

Close Combat

Table of Contents

INTRODUCTION

Close combat is the oldest form of combat known to man. As man progressed, so did his methods of combat. But no matter how technical or scientific warfare becomes, there will always be close combat. When modern weapons fail to stop the opponent, Marines must rely on their close combat skills.

Close combat is at the opposite end of the combative spectrum from self-defense. Self-defense techniques repel an attack. Close combat techniques cause permanent bodily damage to the opponent with every attack and should end in the opponent's death. As Americans, we are conditioned to fight at the intermediate range of close combat. The intermediate range is the distance from which you deliver a punch or kick; e.g., the distance between boxers. Most Marines believe close combat involves punching and kicking the opponent. In reality, most close combat encounters occur in the grappling stage and involve joint manipulation, choking, gouging, and ripping techniques.

This manual addresses fundamentals of close combat, the linear infighting neural-override engagement (LINE) program, bayonet fighting, and weapons of opportunity.

> Fundamentals of close combat address target areas of the body, weapons of the body, stance, falling and rolling techniques, striking and blocking skills, takedowns, and chokeholds.
>
> The LINE program is a learned system of close combat techniques. It develops a Marine's close combat technique to an instinctive level. LINE I addresses wristlocks and counters against chokeholds. LINE II counters against punches and kicks. LINE III addresses unarmed defense against a knife. LINE IV addresses knife fighting. LINE V addresses removal of enemy personnel. LINE VI addresses unarmed defense against a bayonet attack.

Bayonet fighting addresses fighting with an M16A2 rifle.

Weapons of opportunity address fighting with equipment and objects found on the battlefield.

> This manual demonstrates all techniques from the right-hand perspective. However, all techniques can be executed from either side.

FUNDAMENTALS OF CLOSE COMBAT

Target Areas for Unarmed Combat

Close combat's goal is to cause permanent damage to the opponent's body with every technique. To accomplish this, you must know the body's major target areas. Major target areas include the head, neck, torso, groin, and extremities. Target areas are attacked violently and swiftly—there are no second chances. Accessible parts of the opponent's body will vary with each situation and are attacked as presented.

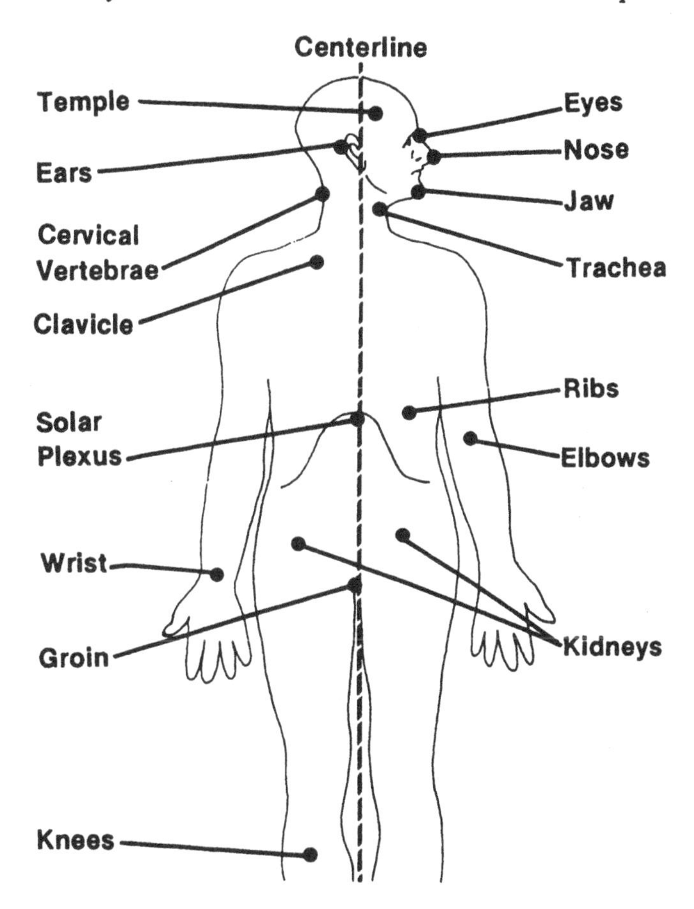

Head. The target areas of the head are the eyes, temples, nose, ears, and jaw. Causing extensive damage to the skull is a primary means of destroying an opponent.

The eyes are soft tissue areas that are not covered by natural protection (i.e., muscle or bone). An attack to the eyes causes the central nervous system to override conscious thought and the opponent involuntarily protects his eyes with his hands, allowing secondary attacks to other target areas.

Powerful attacks to the temple can permanently damage or kill the opponent.

The nose is very sensitive and easily broken. An attack to the nose can cause involuntary watering and closing of the eyes and make the opponent vulnerable to secondary attacks. Through training, individuals can become accustomed to receiving and overcoming attacks to the nose. Therefore, strikes to the nose must be powerfully delivered and immediately followed with a secondary attack.

Attacks to the ear may cause the eardrum to rupture. The fighter must create pressure by cupping his hands and forcefully delivering the strikes.

Forcefully striking the jaw can cause unconsciousness and painful injuries to the teeth, lips, and tongue. However, strikes to the jaw increase your chance of injury. If possible, strike the jaw with a hard object (e.g., helmet, rifle butt, boot heel) to reduce your chance of injury.

Neck. The target areas of the neck are the throat and the back of the neck.

The throat is a soft tissue area and is not covered by natural protection. Damage to the throat causes the trachea to swell and closes the airway which can lead to death.

The back of the neck contains the spinal cord. Attacks to the spinal cord can cause permanent damage and immobilize the opponent.

Torso. The target areas of the torso are clavicle, solar plexus, ribs, and kidneys. During combat, these target areas are usually protected by body armor and combat equipment.

If fractured, the clavicle (or collarbone) can immobilize the opponent's arm.

Attacks to the solar plexus (or center of the chest) can immobilize the opponent by knocking the breath out of him.

Damage to the ribs can immobilize the opponent and cause internal trauma.

Powerful attacks to the kidneys can immobilize, permanently damage, or kill the opponent.

Groin. The groin area is a soft tissue area and is not covered by natural protection. Damage to this area causes the opponent to involuntarily protect the injured area with his hands and legs. The genitals are the main target. A near miss can cause severe pain, contract the lower abdominal muscles, deteriorate the opponent's stance, and cause internal trauma.

Extremities. Typically, an opponent's extremities are encountered before any other major target area. An attack to an extremity (arms and legs) rarely causes death. However, extremities are still important target areas during close combat. The joints of the extremities are the main target areas. Damage to a joint can cause the central nervous system to override conscious thought and immobilize the opponent.

Target Areas for Knife Fighting

Major knife fighting target areas include the head, neck, torso, groin, and extremities. Accessible parts of the opponent's body will vary with each situation. Although there are many insértion and slash points on the human body that can cause permanent damage or death, only the easily accessible target areas are covered.

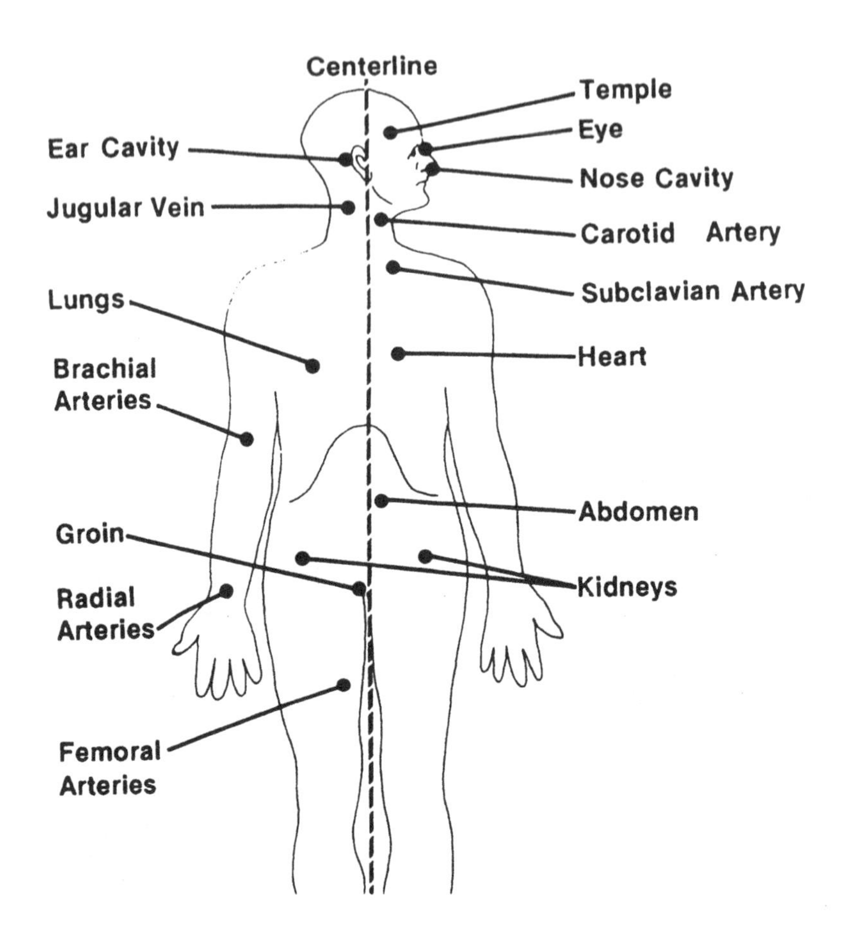

Head. The head area is an excellent target. Typically, slash wounds to the head do not cause death, but they can cause psychological shock and hesitation in the opponent. Knife wounds to the head are beyond the psychological limits of acceptance for most individuals and cause them to hesitate, stop their attack or even quit. Bone surrounding the head can cause the knife blade to deflect and only inflict superficial damage. The main target areas of the head are the temples and the eyes. These areas are protected by a very thin layer of bone and are easily penetrated by a knife blade. Other target areas (i.e., ear, nose, and under the chin) are less accessible and difficult to attack.

Neck. The neck's main target areas are the carotid artery and jugular vein. Precision is not critical when attacking the neck area. The neck area is small and all targets are close together. An attack to the neck area causes extensive damage to the carotid artery; jugular vein; and other arteries, veins, and nerves and can kill the opponent.

Torso. Knife wounds to the torso can disable the opponent or cause death if a vital organ is damaged or the individual goes into traumatic shock. The main target areas of the torso are the subclavian artery, heart, lungs, abdomen, and kidneys. However, these target areas may be inaccessible if the opponent is wearing combat gear (e.g., flak jacket, cartridge belt with canteens).

Overhead stabbing attacks to the subclavian artery can kill the opponent. The subclavian artery is located in the shoulder area and surrounded by the collarbone.

Knife attacks to the heart can kill the opponent. However, the heart is protected by the rib cage. Although the rib cage provides protection, you can damage the heart by inserting a knife blade between the ribs, under the rib cage through the abdomen, or from above the rib cage through the neck.

Knife attacks to the lungs can kill the opponent. However, the lungs are protected by the rib cage. Although the rib cage provides protection, you can damage the lungs by inserting a knifeblade under the rib cage through the abdomen.

The abdomen is an excellent target because of its lack of natural protection. Knife attacks to the abdomen can disable or kill the opponent. If attacking the abdomen, insert the blade and rip across the abdomen to create as large a wound as possible.

Knife attacks to the kidneys can immobilize and incapacitate the opponent. To ensure success, the kidneys should be attacked from the rear. The kidneys are usually protected by combat gear and are difficult to attack.

Groin. The groin area is a soft tissue area and is not covered by natural protection. Knife attacks to the groin area contract the lower abdominal muscles. This causes the opponent to double over in an attempt to protect the injured area. Although a groin wound has the potential to cause death, its main purpose is to immobilize and incapacitate the opponent. Injury to the groin area can also produce shock, fear, and panic.

Extremities. An opponent's extremities (arms and legs) are readily accessible target areas. An attack to the extremities rarely causes death. Although the extremities contain arteries (radial and brachial arteries in the arms and femoral arteries in the upper legs) that can cause death, attacks to the extremities can disable or distract the opponent and make him vulnerable to the main attack.

Personal Weapons

To be successful during close combat, you must know and understand the body's weapons. The body's three main weapon groups are the head, arms, and legs.

Using your body as a weapon increases your chance of injury. Physical damage must be expected during close combat. For example, you may injure your heel while crushing your opponent's skull.

Head Movement

Striking with your head or biting increases your chance of injury. Do not use your head as a striking weapon unless a helmet is worn. Although these techniques are not recommended, you must use all techniques at your disposal to survive an encounter.

Arm Movement

Your hand is your arm's most versatile weapon. The knuckles, cutting edge of the hand, heel of the palm, and fingers can be used as weapons.

Your hand can be balled into a fist and your knuckles can be used to strike the opponent. Striking with your fist often causes injuries to your hand and is not recommended as a primary method of attack. If using your knuckles to attack the opponent, direct the blow to soft tissue areas (i.e., eyes, throat, groin) to reduce your chance of injury.

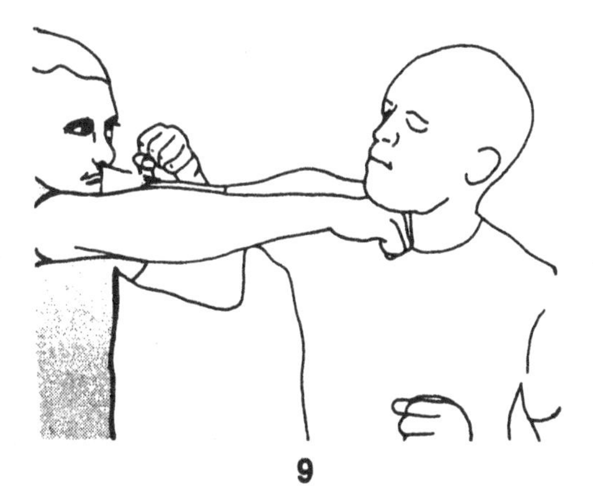

The cutting edge of your hand can be used as a weapon to strike the soft tissue areas of the eyes and throat.

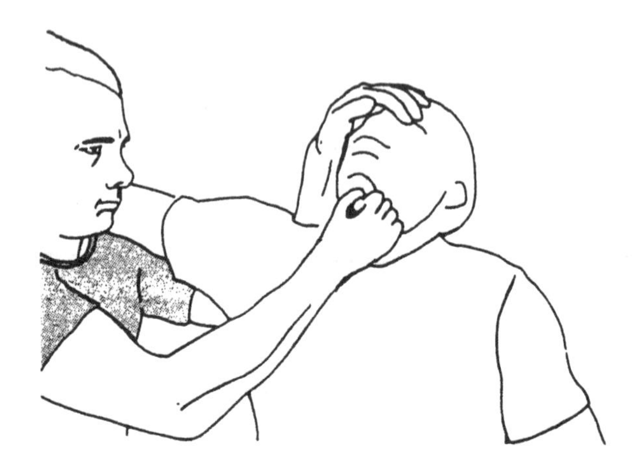

The heel of your palm can strike, parry, and block the opponent.

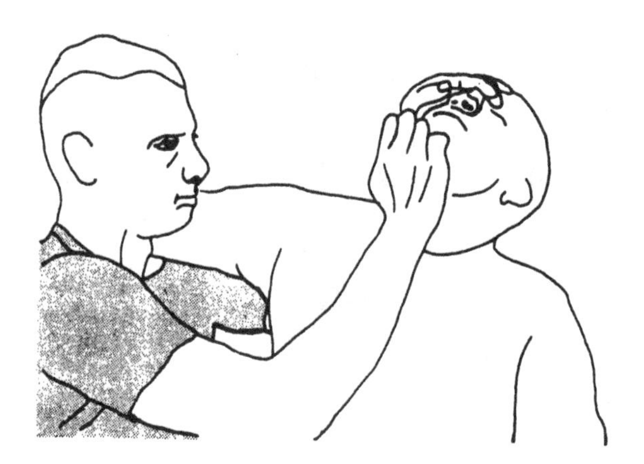

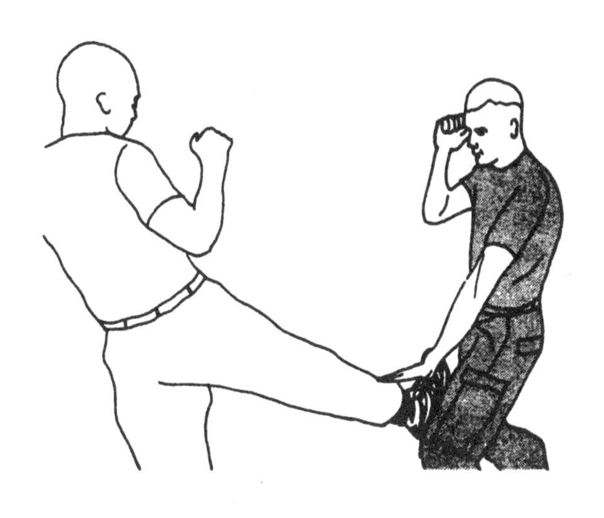

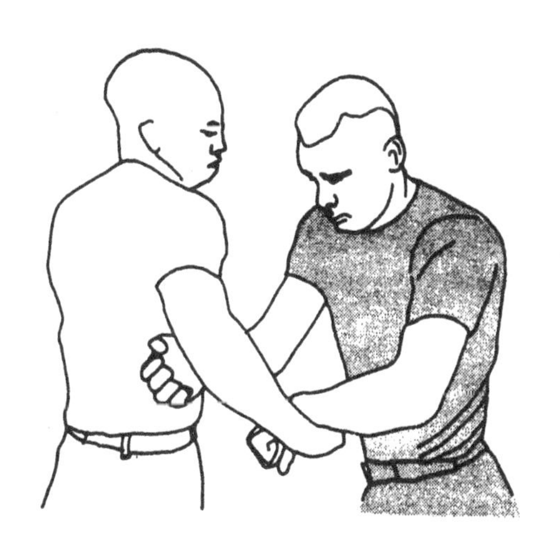

Your fingers can gouge, rip, and tear the soft tissue areas of the eyes, throat, and groin.

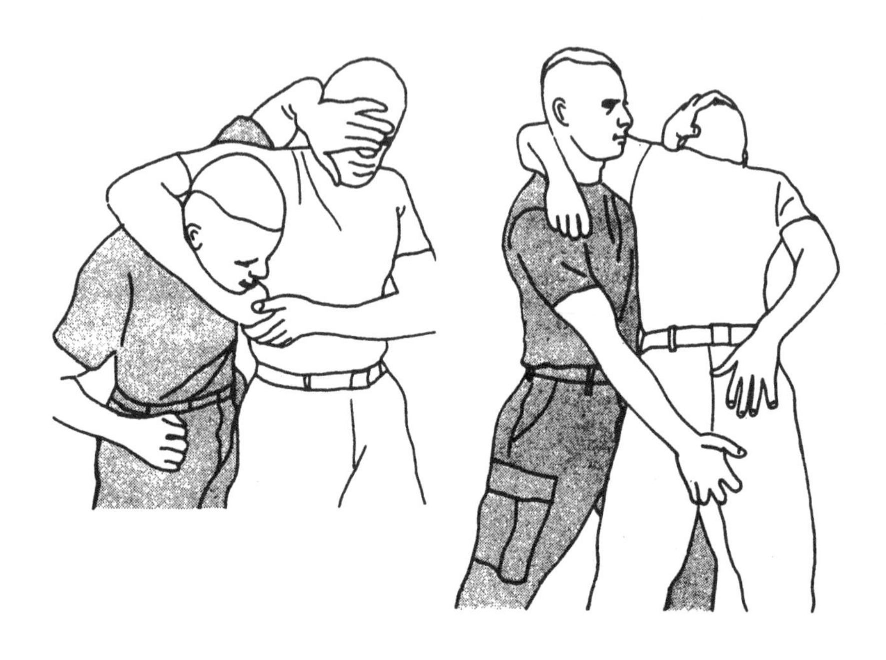

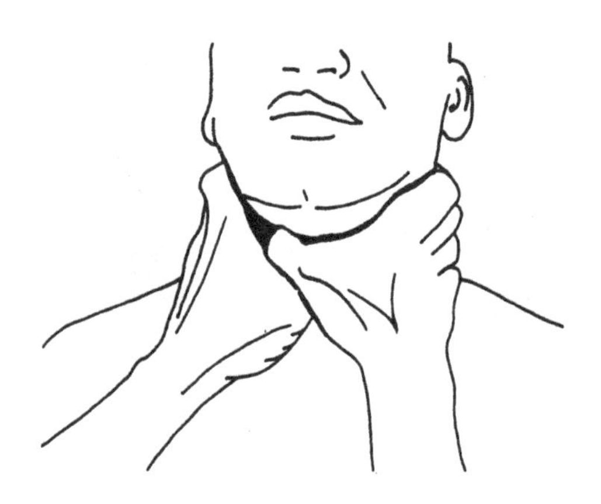

Your forearm is extremely important during a defensive posture. Your forearm can block or parry attacks . . .

. . . or it can strike and break the opponent's elbow. Using your forearm as a striking weapon helps prevent injuries to your fist and fingers.

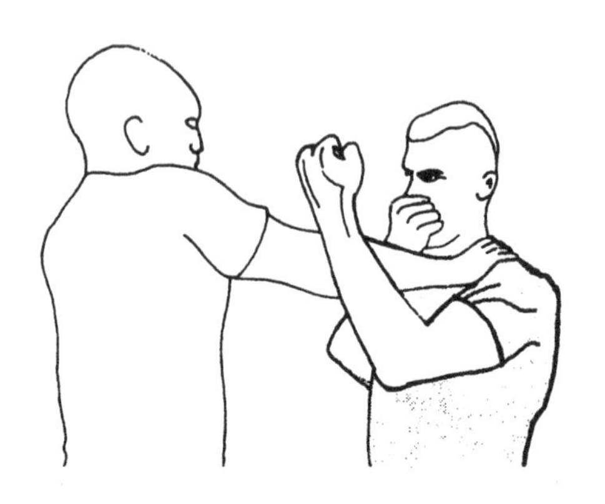

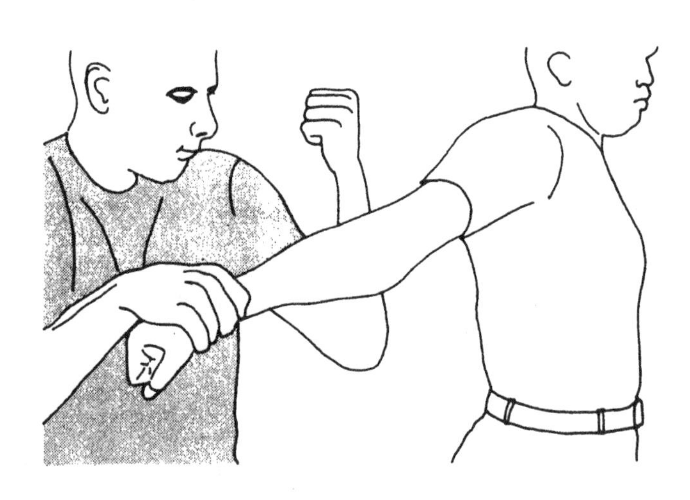

Your elbow is a devastating striking weapon because it delivers a powerful blow within a short distance. This makes your elbow an excellent striking weapon during the grappling stage of close combat.

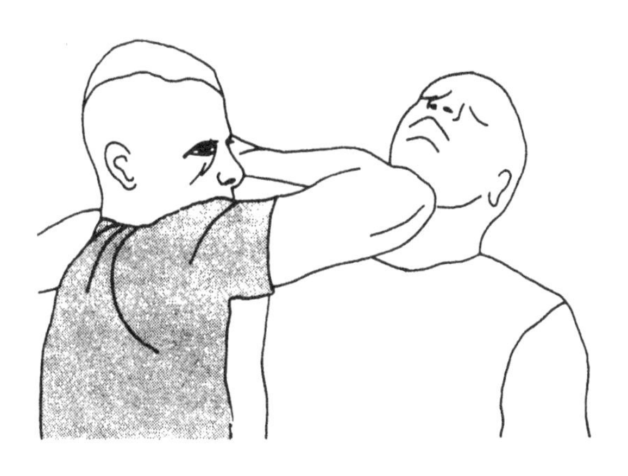

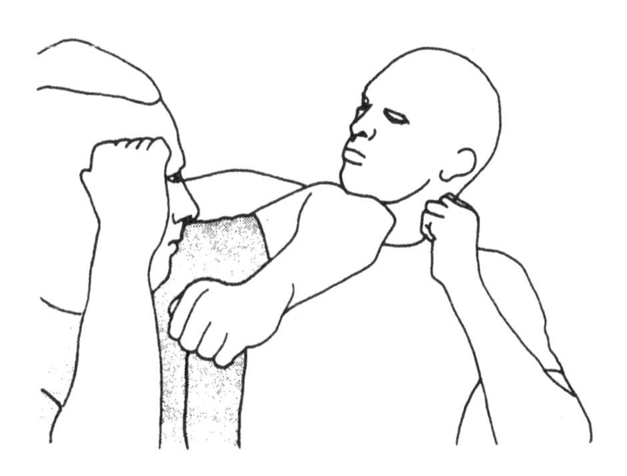

Leg Movement

Your legs are more powerful than your arms or head and are less prone to injury. Your foot is protected by your boot and is the preferred choice for striking the opponent. The toe, ball of the foot, instep, bottom of the heel, and cutting edge of the heel can be used to strike the opponent. The knee is extremely effective during infighting.

Your toe is recommended for striking only while wearing boots. You can use your toe,

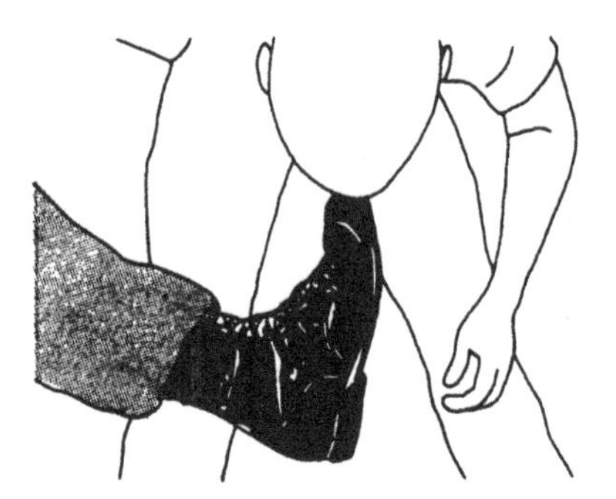

the ball of your foot,

your instep,

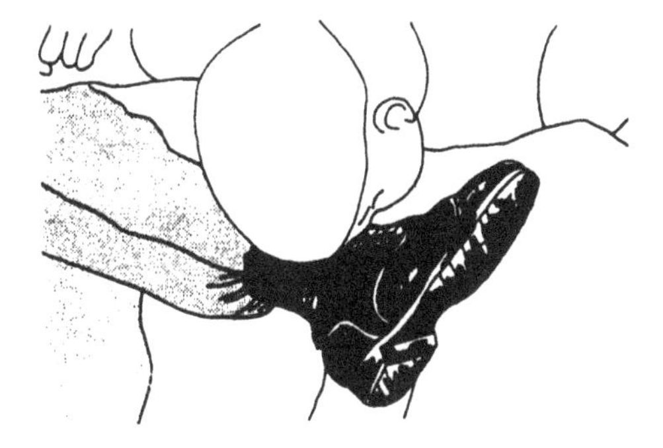

the bottom of your heel,

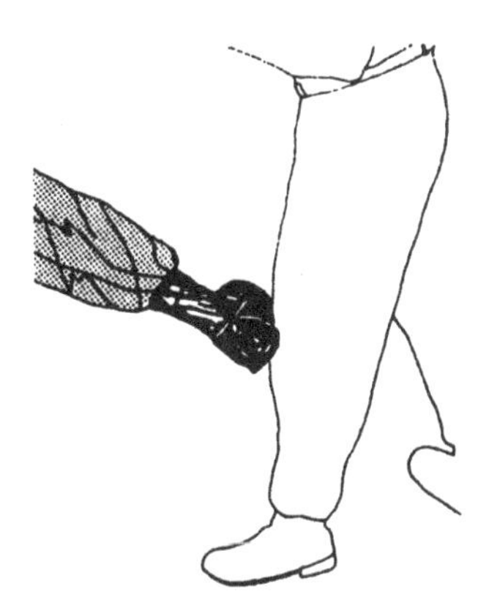

or the cutting edge of your heel to strike the opponent.

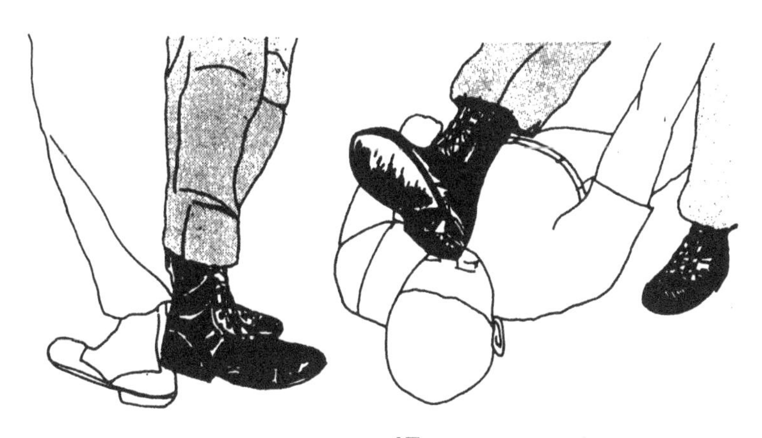

You can also use the knee strike to deliver a devastating secondary attack to the opponent's face or groin.

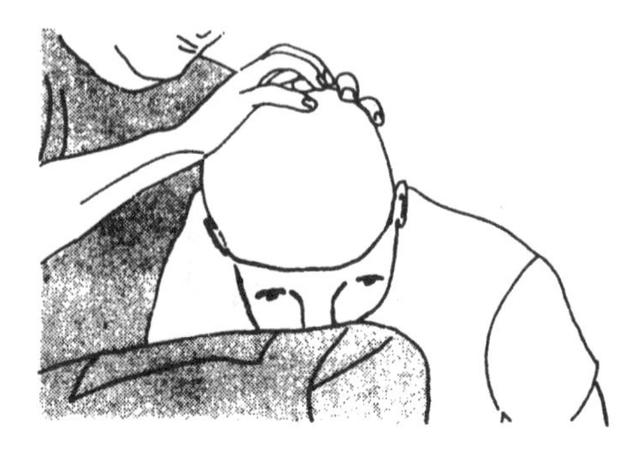

The Basic Warrior Stance

It is difficult to develop effective close combat skills without a solid stance. A solid stance is critical to all phases of close combat. The basic warrior stance provides the foundation for all movements and techniques and must be mastered by all Marines. Perform the following steps to attain the basic warrior stance.

- Place your feet shoulder-width apart with the toe of your rear foot in line with the heel of your front foot. Feet should point 45 degrees from the direction of attack.
- Bend slightly at your knees.
- Distribute your body weight evenly on both legs.
- Bend your elbows to form 45 degree angles.
- Hold your arms high enough to protect your face. Arms should not block your vision.

- Keep your elbows close to your body to protect your ribs.
- Curl your fingers into a fist. Do not clench your fist. Clenching contracts your forearm muscles and detracts from hand speed and reaction time.
- Tuck your chin down to take advantage of the natural protection provided by your shoulders.

You must be able to assume the basic warrior stance instinctively and move in all directions while maintaining the basic warrior stance. During movement, your legs or feet should not be crossed, your upper body should remain in the basic warrior stance, and your knees bend deeper than normal. Movement is executed through your legs. Do not bend your waist to aid in movement. If possible, use hand movements (feints, strikes) to conceal the movement of your legs and feet.

Forward Movement

Quickly slide your lead foot forward approximately 12-15 inches. As soon as your lead foot is in place, quickly move your rear foot forward to return to the basic warrior stance.

Rear Movement

To move to the rear, execute the forward movement in reverse. Quickly slide your rear foot to the rear approximately 12-15 inches. As soon as your rear foot is in place, quickly bring your lead foot to the rear to return to the basic warrior stance.

Changing Directions

Sometimes you must change direction in order to face the opponent. To change direction, quickly turn your head to the new direction, push off with your lead leg, and quickly step in the new direction while pivoting on the ball of your rear foot. Upon completion of the movement, you should be in the basic warrior stance.

Note

It is important that your head turn quickly to the new direction. The faster your head turns, the faster your body can move and the quicker you attain visual contact with your opponent.

Breaking a Fall

There may be times during an encounter in which you lose your balance or are thrown by the opponent to the ground. Your body's muscles can be used to protect vital organs and bones from injury. By using the large muscle groups (back, thighs, buttocks) to cushion the impact of a fall and to maintain motion after hitting the ground, you can avoid serious injury and immobilization.

Try to use the momentum of a fall to maintain motion. It is important to remain in the basic warrior stance, even while falling or being thrown, and ensure that your head is tucked tightly into your chest. By maintaining your stance, using your large muscle groups as a cushion, and using the momentum of the fall to maintain motion and reduce the force of the impact, you reduce your chances of serious injury and increase your chances of survival.

Note

Do not throw your arm out to break your fall. This may work well on a mat or in an area with no debris, but in a combat environment you will not know what is on the ground. If you extend your arm and strike something hard at the elbow, you can effectively take yourself out of the fight.

Shoulder Roll

You can use the momentum of a fall to execute a shoulder roll. During a shoulder roll, the large muscle group of your upper back absorbs the impact of the fall rather than your neck and spinal column. To execute the shoulder roll from the basic warrior stance –

- Tuck your chin and rear shoulder in.
- Keep your arms close to your sides.
- Roll forward.
- Continue to roll until you are standing upright.
- Resume the basic warrior stance.

Practice the shoulder roll while unarmed and armed with a rifle. The following pages illustrate an unarmed and armed shoulder roll.

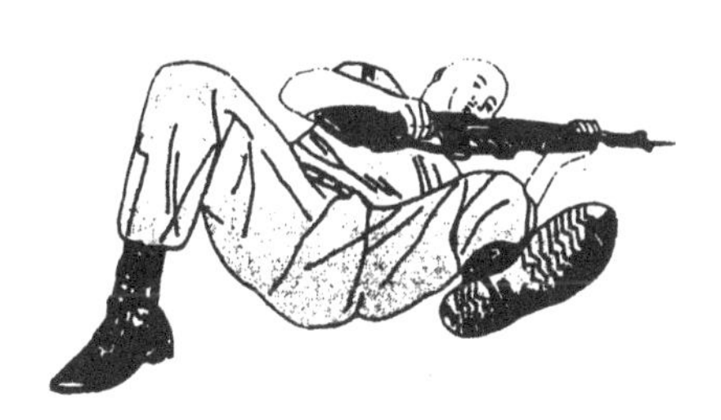

Offensive Skills

During self-defense, the goal is to repel the opponent. During close combat, the goal is to destroy the opponent while maintaining control. Control reduces the opponent's opportunity to retaliate, ensures that the opponent remains stationary, and allows you to deliver accurate strikes.

Striking techniques should be practiced until they become instinctive and can be applied with speed and force. Striking techniques form the basis for armed techniques such as knife and bayonet fighting.

A perfectly executed strike may not be enough to eliminate an opponent. To ensure that the opponent is destroyed, **deliver strikes violently, swiftly, and repeatedly.** To reduce your chance of injury, direct hand strikes to a soft tissue area (eyes, throat, and groin).

Lead Hand Punch

The lead hand punch is a snapping straight punch executed by your forward or lead hand. The knuckles of your hand make contact with the opponent. This technique conceals movement and allows you to close with the opponent. Lead hand punches should strike soft tissue areas if possible.

Rear Hand Punch

The rear hand punch is a powerful straight punch executed by your back or rear hand. The knuckles of your hand make contact with the opponent. The punch's power comes from your rear leg and forceful rotation of your hips and shoulders. Your center line (eyes, throat, groin) becomes exposed as your hips rotate toward the target; therefore, a strike with the lead hand should precede this technique.

Forearm Strike

The forearm strike is effective against a variety of targets – especially the elbow. The forearm strike is executed by your arm. The inside or outside of your forearm makes contact with the opponent. To be successful, your nonstriking arm must trap and immobilize the opponent's joint.

If delivered from close to your torso, the forearm strike carries the weight and power of your entire upper body. This is critical if you are physically weak or exhausted.

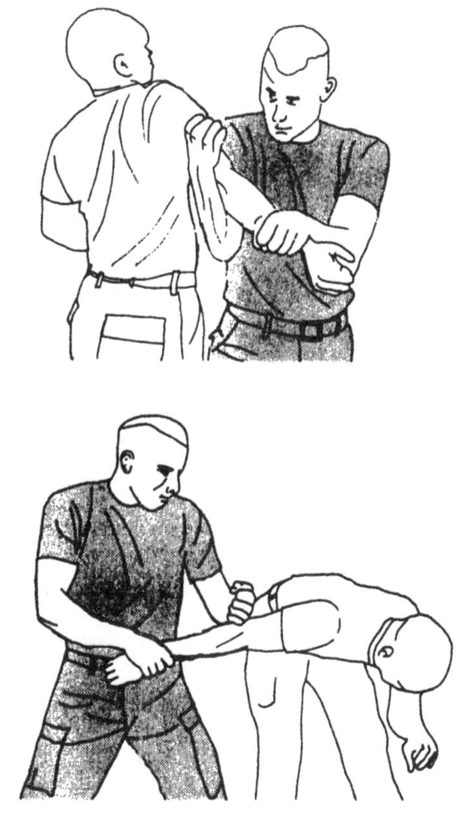

Elbow Strike

The elbow strike is executed by your arm. The forearm side of your elbow or the tip of your elbow makes contact with the opponent. The strike's power comes from the forceful rotation and drive of your hips and shoulders. The elbow strike can be thrown by either arm in a striking or jabbing motion. The rear arm is preferred because it allows the lead arm to immobilize the opponent.

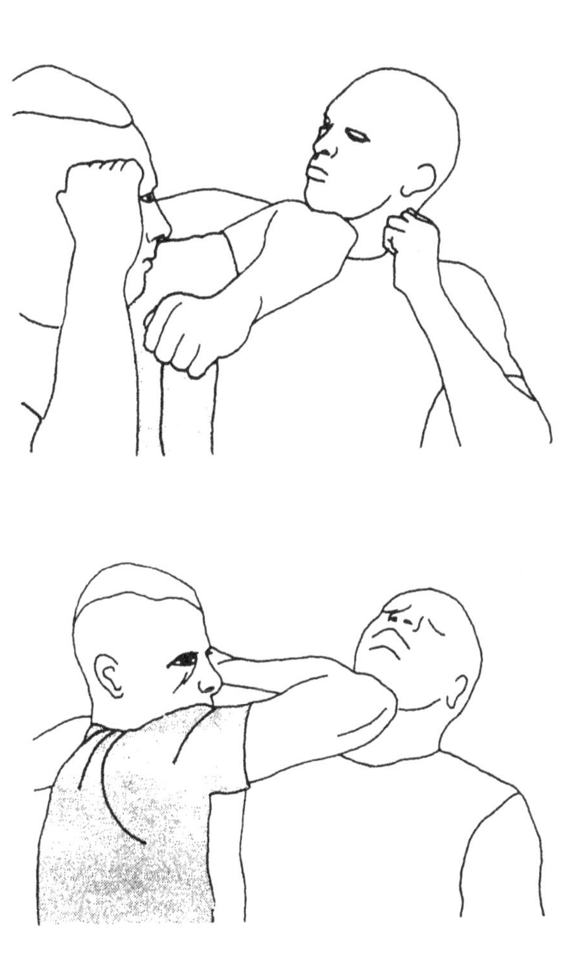

Knifehand Strike

The knifehand strike resembles the chopping motion of a knife. The knifehand strike is executed by your rear hand. The outside edge of your hand between the joint of the small finger and the wrist makes contact with the opponent. The main purpose of this strike is to destroy the opponent. The preferred target area for a knife-hand strike is the throat. To be successful, your nonstriking hand must control the opponent.

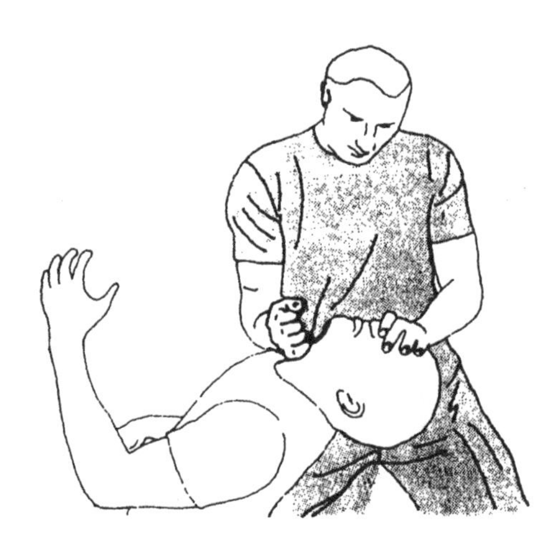

Knee Strike

The knee strike is effective during infighting. The knee strike is executed by your leg (either horizontally or vertically). The top of your knee makes contact with the opponent. The strike's power comes from forcefully lifting your thighs and pivoting the hips.

To be successful, a control technique is used with the knee strike to immobilize the opponent.

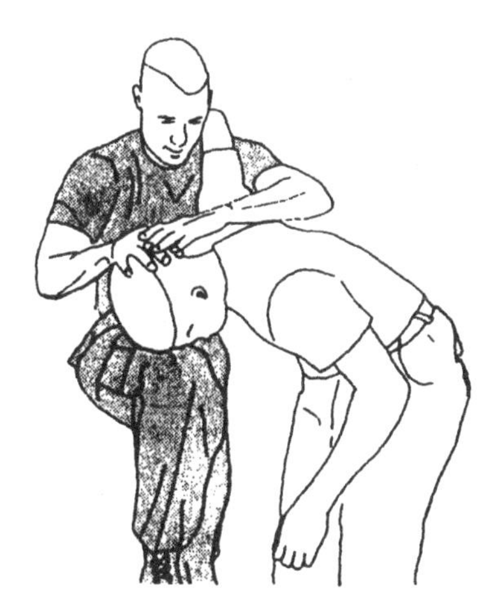

Kicks

Kicks can stop an opponent's attack or create an opening in his defenses. Kicks used in close combat must be simple and easily executed under combat conditions (e.g., combat gear, fatigue, darkness). Kicks delivered above the waist expose your groin and decrease balance. Kicks delivered to the opponent's waist or below his waist should immobilize the opponent—not merely drive him away.

Lead Leg Front Kick. The lead leg front kick is executed by your lead leg. The toe of your boot or the ball of your foot make contact with the opponent. Execute the lead leg front kick by quickly raising your lead knee and snapping your leg toward the target area (e.g., groin, knee). After contact with the target area, quickly return your leg to the basic warrior stance. The speed of this kick reduces your chances of injury and reduces possible counterattacks by the opponent.

Lead Leg Side Kick. The lead leg side kick is executed by your lead leg. The edge and bottom of your heel and sole of your boot make contact with the opponent. Execute the lead leg side kick by quickly raising your lead knee and snapping your leg while rotating your hip toward the target area (e.g., knee, ankle). After contact with the target area, quickly return your leg to the basic warrior stance. The side kick allows your hips to remain closed and protects your groin.

Rear Leg Front Kick. The rear leg front kick resembles the punting of a football. The rear leg front kick is powerful and can cause extensive damage. The rear leg front kick is executed by swinging your entire leg upward. Your instep and toe make contact with the opponent. The kick's power comes from your hips and thighs. The target area is the opponent's face or throat. Do not deliver the rear leg front kick above your waist. To be effective, the opponent must be controlled and immobilized. Use joint manipulation to control the opponent and to bring his head below his waist.

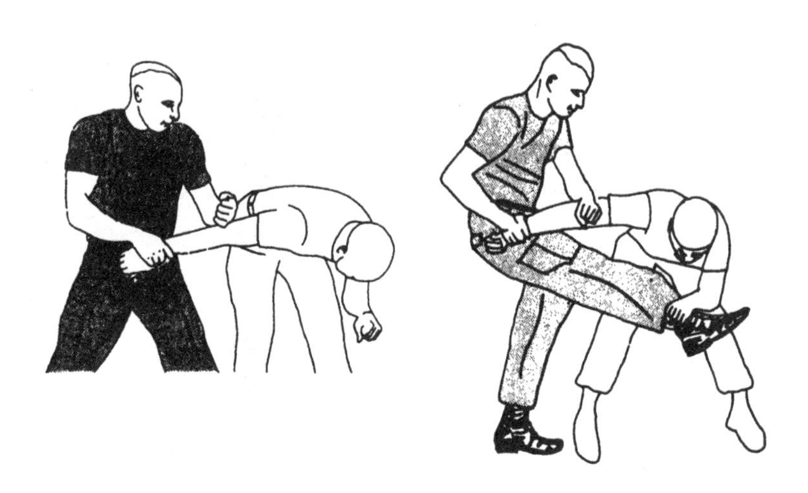

Heel Stomp. The heel stomp is an excellent finishing technique. If executed correctly, the heel stomp can damage any part of the opponent's anatomy it strikes. The heel stomp is executed by quickly swinging your rear leg at least waist high over the target and driving the cutting edge of your boot heel straight down into the target area with as much speed and force as possible. The cutting edge of your boot heel makes contact with the opponent. The higher your leg is raised, the greater the velocity and force. The target areas are the skull and neck.

Leg Sweep

The leg sweep is used to take down an opponent. Before attempting a takedown, the opponent's injuries should prevent retaliation in order to ensure that the opponent goes to the ground. To execute the leg sweep, maintain balance while swiftly—

- Raising your rear leg as high as possible to the rear of the opponent.
- Driving your rear leg downward forcefully.
- Striking the opponent's achilles tendon.
- Completing the momentum begun by your leg.

The cutting edge of your boot heel makes contact with the opponent. To be successful, maintain control of the opponent throughout all phases of the takedown. Takedowns executed without control of the opponent risks completing the destruction of the opponent.

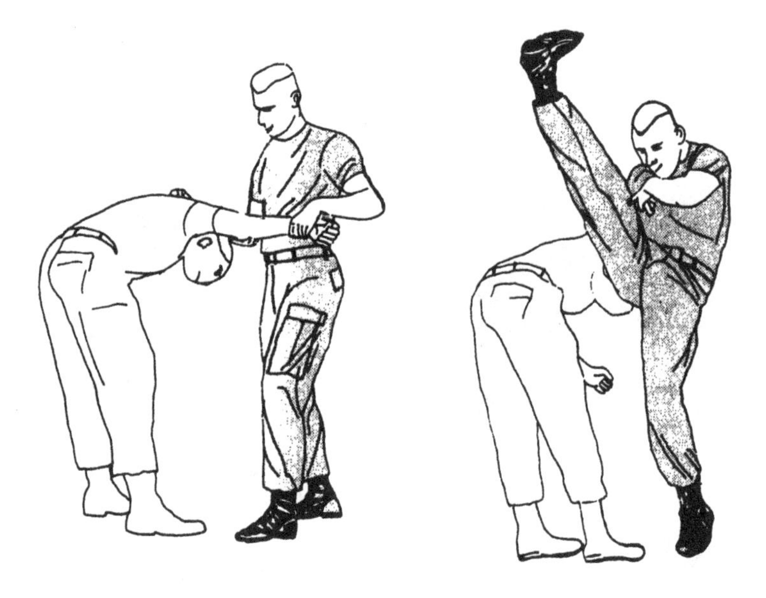

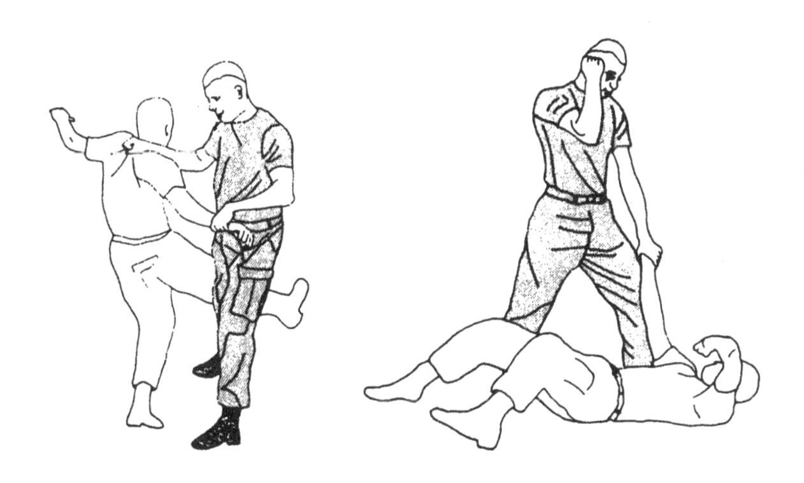

Chokes

The choke closes the airway and causes death by asphyxiation or cutting off the blood flow to the brain. Choking is not as effective a finishing technique as striking. However, you should be able to execute a choke **swiftly and forcefully.** While executing the choke, your vital areas are vulnerable to counterattacks. You must damage the opponent's windpipe before he can counterattack. For maximum leverage and to prevent counterattacks, press your arms and body against the opponent's body while executing a choke. To execute a choke from the front–

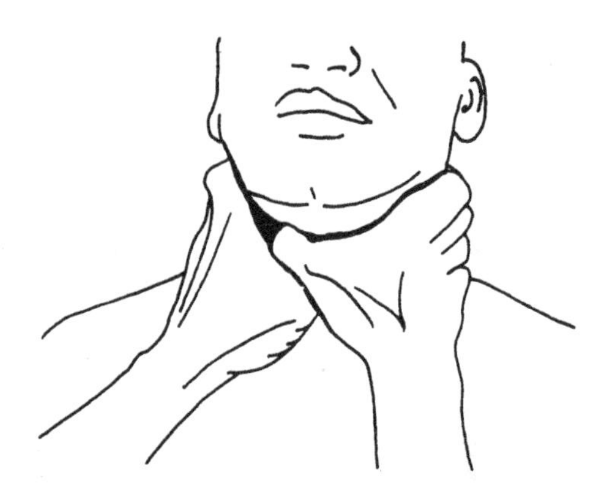

To execute a choke from the back–

Defensive Skills

The goal of defensive skills is not only to defend against an attack, but to put yourself in position to counterattack. Defensive movements should not break the balance of the basic warrior stance.

While defending against an incoming attack, the defending arm (lead arm) moves out of position only enough to engage the attack. The nondefending arm (rear arm) maintains its position. Because the lead hand is closest to the opponent, it assumes most of the defensive duties. The rear hand backs up the lead hand and blocks incoming attacks to the right side of the head and torso.

It is best to block or parry on an angle so you do not directly oppose the full force of an attack. Blocking or parrying lessens the force of impact, allows better opportunity for a counterattack, and protects the defensive zone. To ensure success, blocking moves are executed with as much speed and force as possible. The defensive zone is the area an attack must enter to cause damage. Occasionally, a fighter breaks the basic warrior stance and reacts to an attack delivered from outside the defensive zone. By doing this, the fighter is unable to engage the opponent and has exposed himself to a secondary attack. Do not defend against attacks that are delivered outside the defensive zone.

Blocking

High Block. The high block defends against overhead attacks. To execute the high block—

- Close your hand to prevent finger injuries.
- Snap your forearm up.
- Clear your head enough to engage the attack—do not over-extend.
- Bend your elbow.
- Apply tension to both your elbow and shoulder.

The opponent should strike the outside of your forearm.

Low Block. The low block defends against attacks to your mid-section and groin. To execute the low block—

- Close your hand to protect your fingers.

- Snap your forearm down the front of your body to engage the attack.
- Apply tension to your elbow and shoulder.

The opponent should strike the outside of your forearm.

Outside Block. The outside block defends against attacks directed at your upper body from the outside and straight-in attacks directed at your upper body. To execute the outside block—

- Close your hand to protect your fingers.
- Snap your blocking arm to the outside of your body.
- Engage the attack.
- Ensure that the attack does not drive your defending arm into your body or head.
- Apply tension to your elbow and shoulder.

The opponent should strike the outside of your forearm.

Inside Block. The inside block defends against straight-in attacks directed at your upper body. To execute the inside block—

- Close your hand to protect your fingers.

- Snap your forearm toward the inside of your body.

- Apply tension to your elbow and shoulder.

The opponent should strike the inside of your forearm.

Leg Block. The leg block defends against low-line kicks to your groin and the joints of your lead leg. By countering an opponent's low-line kick with a leg block, your defensive posture is not compromised. To execute the leg block, raise the knee of your lead leg. The opponent should strike your lead leg.

Defensive Prone Position

The defensive prone position defends against attacks while you are on the ground and unable to regain the basic warrior stance. To assume the defensive prone position—

- Position your body on its side.
- Tuck your rear leg under your body for stability.
- Place your rear arm under your body with the palm of your hand on the ground.
- Move your rear arm to the right or left while pivoting on your hip to move your body.
- Keep your lead arm in the basic warrior position.
- Position your hand and forearm to protect your head and neck area.
- Position your bicep/tricep area to protect the ribs.
- Lift and cock your lead leg to protect your groin and strike the opponent's ankle, shin, knee, and groin if necessary.

Make every attempt to regain your footing and resume the basic warrior stance.

THE LINE SYSTEM

The LINE system is the heart of the Marine Corps' close combat training system. The LINE system teaches Marines specific movements and techniques. These movements and techniques identify—

- Application of movements and techniques during close combat situations.
- Response during the grappling stage.
- Response during infighting.
- Damage to the opponent's anatomy that causes his central nervous system to override conscious thought in order to immobilize and control the opponent.
- Elimination of the opponent once control is achieved.

Once the LINE system of movement and techniques has been mastered, Marines should be able to apply these movements and techniques instinctively.

The basic LINE system is divided into six parts. Each part builds upon and reinforces techniques learned previously.

- LINE I teaches the basic offensive, defensive, and takedown techniques used in the grappling stage of close combat. LINE I techniques are the basis for LINES II-IV.
- LINE II teaches techniques used during the intermediate range (punching and kicking) of close combat.
- LINE III teaches techniques used during unarmed defense against a knife attack.
- LINE IV teaches techniques of fighting with a knife when attacked by an opponent armed with a knife.

- LINE V teaches techniques used during removal of enemy personnel.

- LINE VI teaches techniques used during unarmed defense against a bayonet attack. LINE VI is discussed in depth on pages 155 through 178.

Continuously practice the LINE system of techniques. The techniques should become instinctive, conditioned actions. Effective training requires that techniques be executed **swiftly and violently.** Use alternate striking areas and slight modifications to safely allow violent contact during training.

To simulate an eye gouge, grab the opponent's forehead just above his eyes.

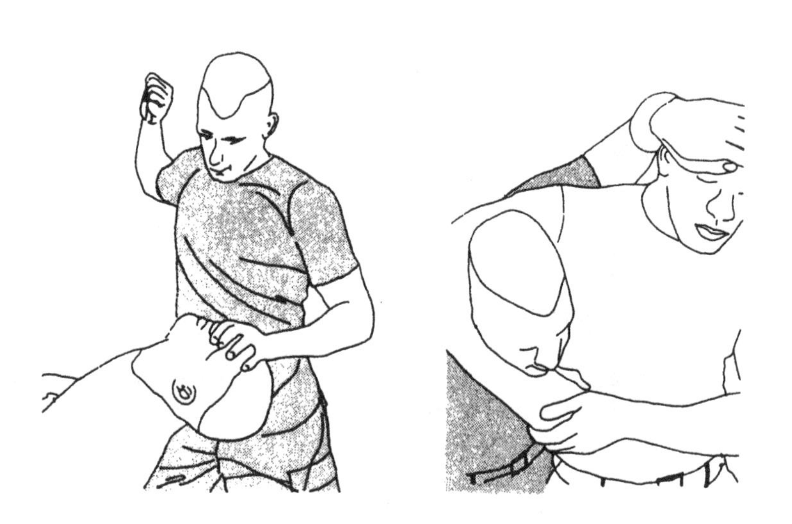

To simulate a groin strike, grab the inside of the opponent's thigh.

To absorb a kick to the head, quickly bring your free arm up in front of your face. Apply tension to your arm and absorb the impact of the kick with your forearm. Practice this defensive movement as an instinctive action.

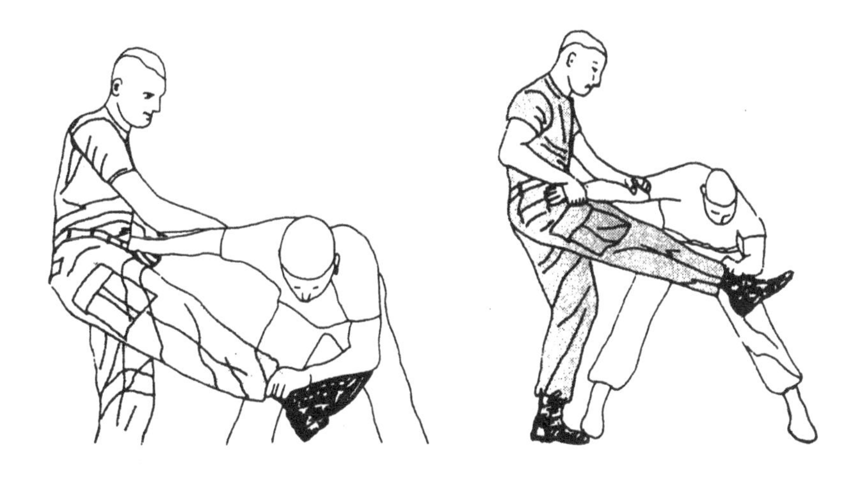

To simulate an elbow strike, bend your arm slightly. Apply tension to absorb the simulated elbow strike.

To simulate a heel stomp, your heel should hit the ground approximately 12 inches from the opponent's head. The opponent should hold his free arm in front of his head for added protection.

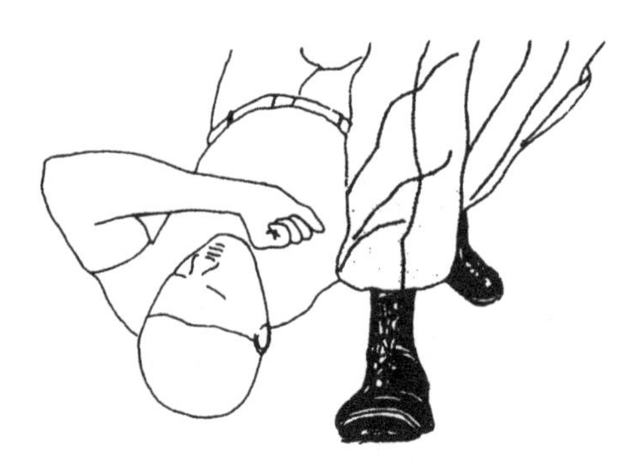

LINE I

Wristlocks and Counters Against Chokes and Headlocks. LINE I teaches the basic techniques used in the grappling stage of close combat. Control techniques alone do not cause death. If applied with force, control techniques can cause damage to the opponent's joints and allow you to gain and maintain control of the opponent while reducing your chance of risk.

Wristlock

The wristlock is a joint manipulation technique used to control the opponent and cause permanent damage to the wrist. To execute the wristlock—

Note

The photos illustrate the fighter defending a lapel grab.

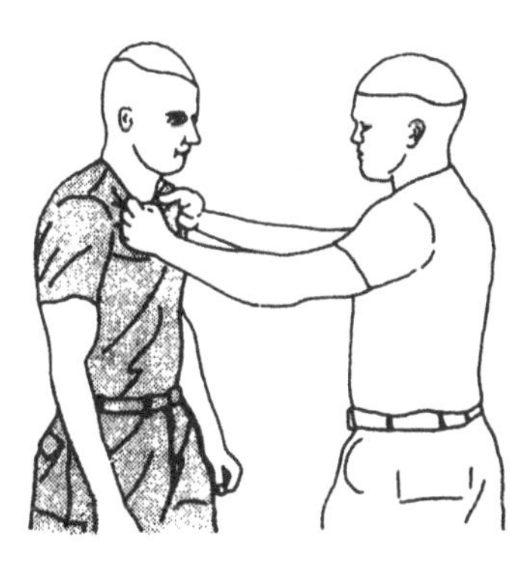

- Reach over the opponent's arm and quickly grasp the opponent's hand.
- Place your thumb in the middle of the back of the opponent's hand.

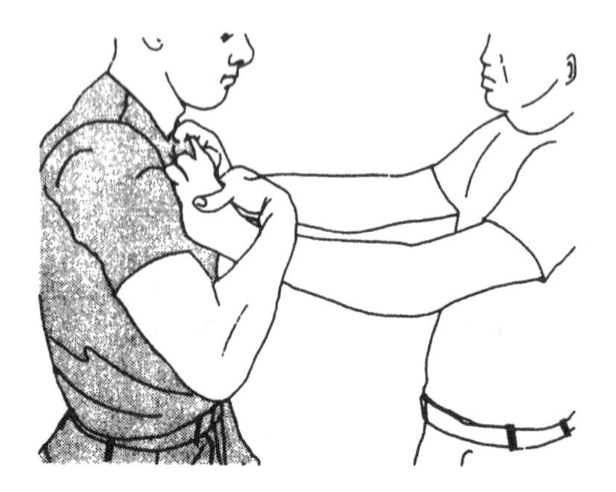

- Wrap your fingers around the opponent's hand beneath his thumb.
- Turn the opponent's hand forcefully upward until the palm is vertical to the ground.

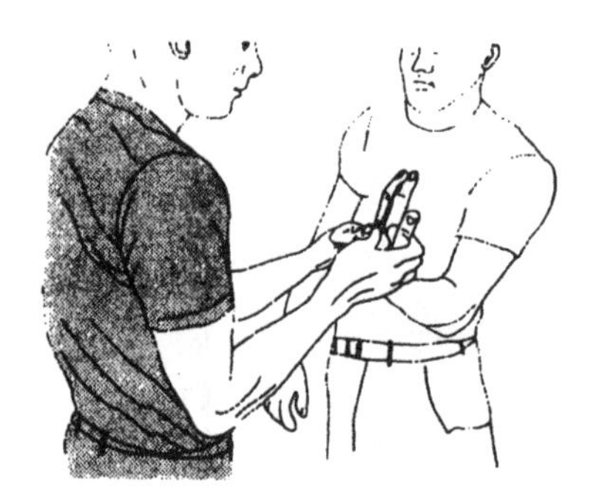

- Grasp the opponent's hand with your free hand, place your thumbs together, and wrap your fingers around the opponent's hand beneath his little finger to provide added leverage.

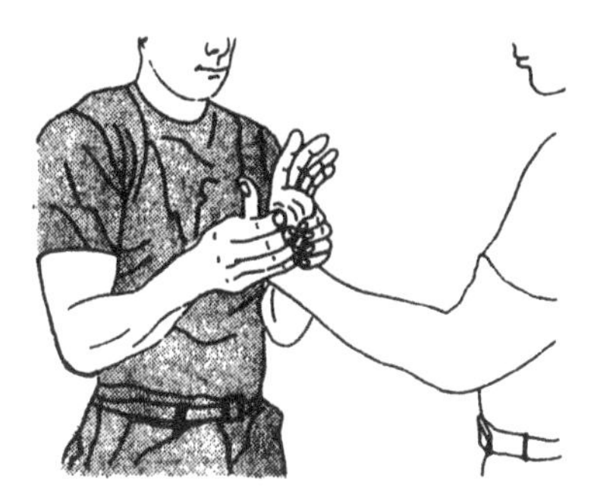

- Push the opponent's hand at an angle to the outside of the elbow.

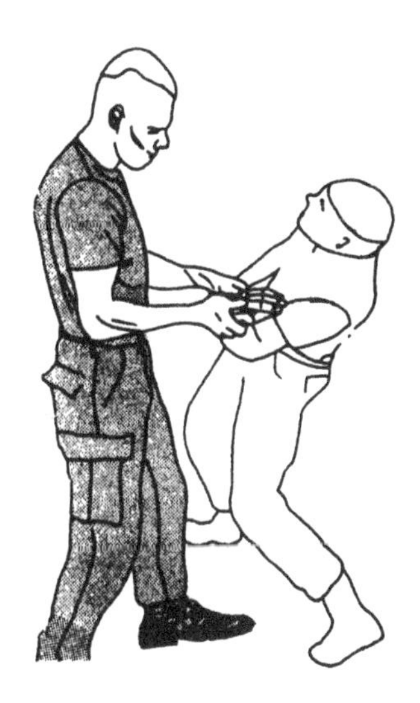

- Push downward until the opponent is on the ground.

- Use your knee to lock the opponent's fully extended elbow while maintaining pressure to the wristlock. This maintains control of the opponent.

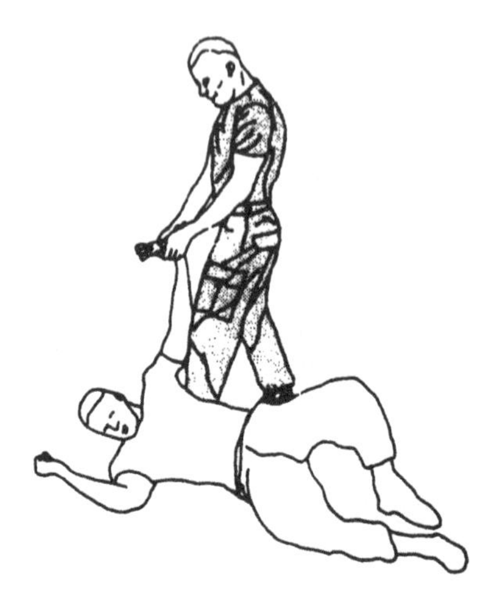

During close combat training, apply the wristlock with smooth, consistent pressure. During close combat, apply the wristlock with a forceful snapping motion.

Reverse Wristlock

The reverse wristlock is a variation of the wristlock. To execute the reverse wristlock—

Note

The photos illustrate the fighter defending a lapel grab.

- Reach over the opponent's arms and quickly grasp the opponent's hand.

- Place your thumb in the middle of the back of the opponent's hand.
- Wrap your fingers around the opponent's hand beneath his little finger.

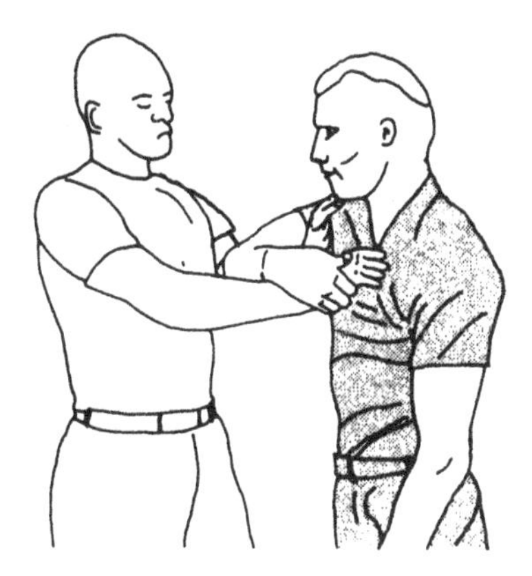

- Turn the opponent's hand forcefully to the inside with his palm vertical to the ground.

- Grasp the opponent's hand with your free hand, place your thumbs together, and wrap your fingers around the opponent's hand beneath his little finger to provide added leverage.

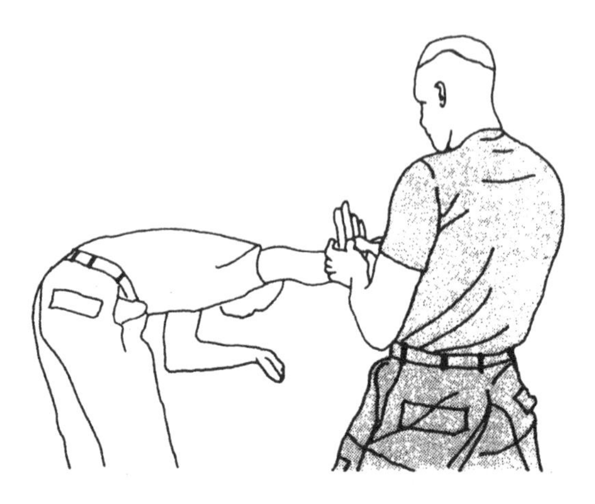

- Apply pressure while driving the opponent's hand forcefully toward his shoulder.

- Push downward until the opponent is on the ground.

- Use your knee to lock the opponent's fully extended elbow while maintaining pressure to the wristlock. This maintains control of the opponent.

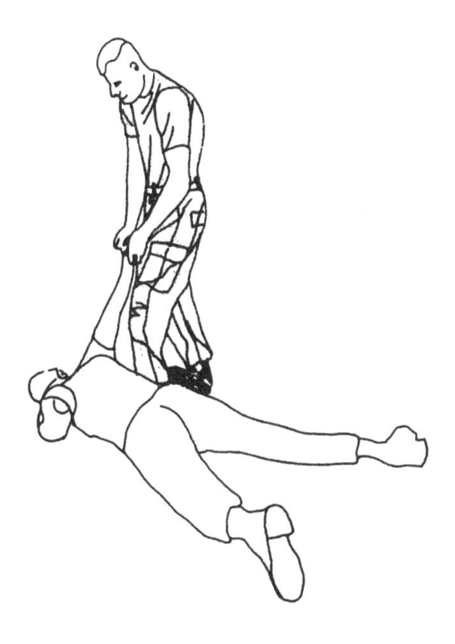

Defense for a Front Choke

If the opponent attempts a choke from the front, forcefully deliver forearm strikes to the opponent's right arm to damage the elbow. To defend against a front choke—

- Strike the opponent's inside right wrist with your right forearm while striking the outside of the opponent's right elbow with your left forearm.

- Step back quickly with your right foot while maintaining pressure to the damaged elbow.

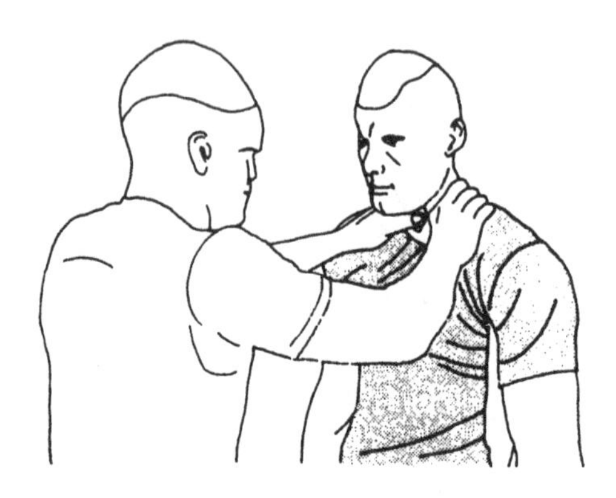

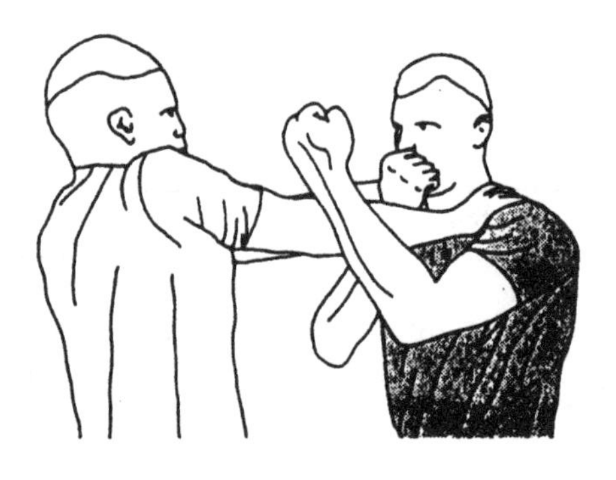

- Grab the opponent's wrist with your right hand and drive your left forearm into the injured elbow causing the opponent to bend at the waist.

- Execute a rear leg front kick to the opponent's face.

- Switch grips quickly by—
    - Grabbing the opponent by the back of the neck with your right hand.
    - Grabbing the opponent's wrist with your left hand.
    - Maintaining control and body contact with the opponent while changing grips.
    - Rotating your hips to face the rear of the opponent and positioning yourself for the leg sweep.

- Execute the leg sweep by striking the opponent's achilles tendon with the cutting edge of your heel and driving him to the ground.

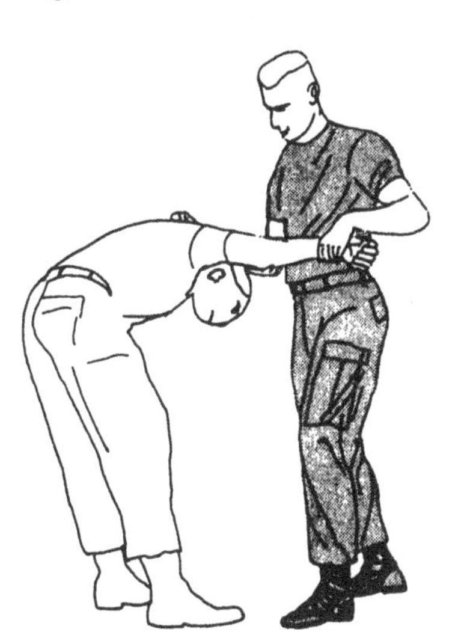

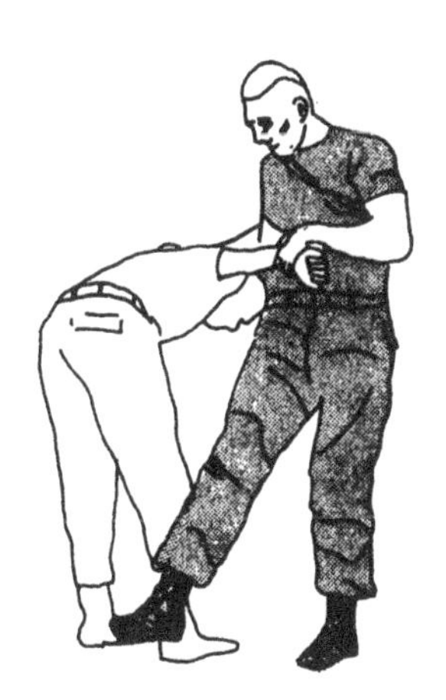

- Return your right hand to the basic warrior position as the opponent falls.
- Keep your body erect while maintaining control of the opponent's arm.

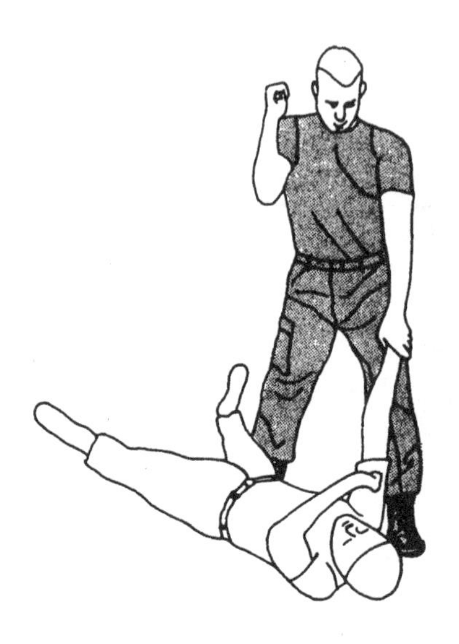

- Execute the heel stomp **swiftly and violently** to the opponent's head as a finishing technique.

Defense for a Rear Choke

If the opponent attempts a choke from the rear, swiftly and forcefully execute an open hand groin strike to loosen the choke. To defend against a rear choke—

- Use the palm of your left hand to strike the opponent's groin and grip the opponent's right wrist with your right hand.

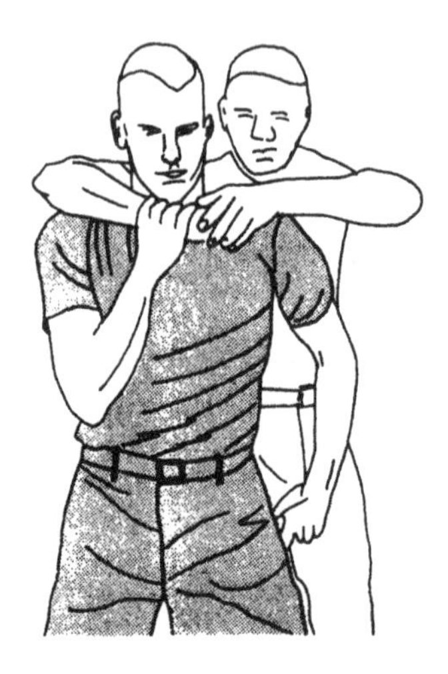

- Grab, squeeze, and twist the opponent's groin violently and forcefully.
- Sidestep quickly to the right under the opponent's right arm as the chokehold loosens.

- Maintain a firm grip on the opponent's right wrist and violently pull his arm to a fully extended and locked position in front of him.

- Deliver a powerful left forearm strike to the opponent's elbow once the elbow is in a fully extended and locked position. This damages the elbow and drives the opponent's upper body down.

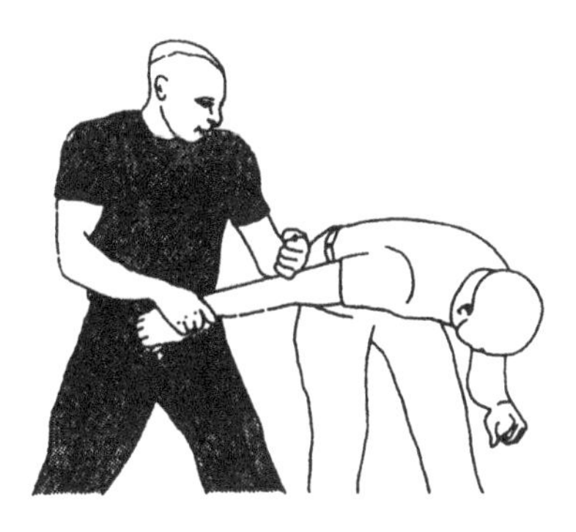

- Maintain control of the opponent's arm and apply pressure to his injured elbow.
- Execute a rear leg front kick to the opponent's face.

- Grab the opponent behind the neck, rotate your hips, and execute a leg sweep taking the opponent to the ground.

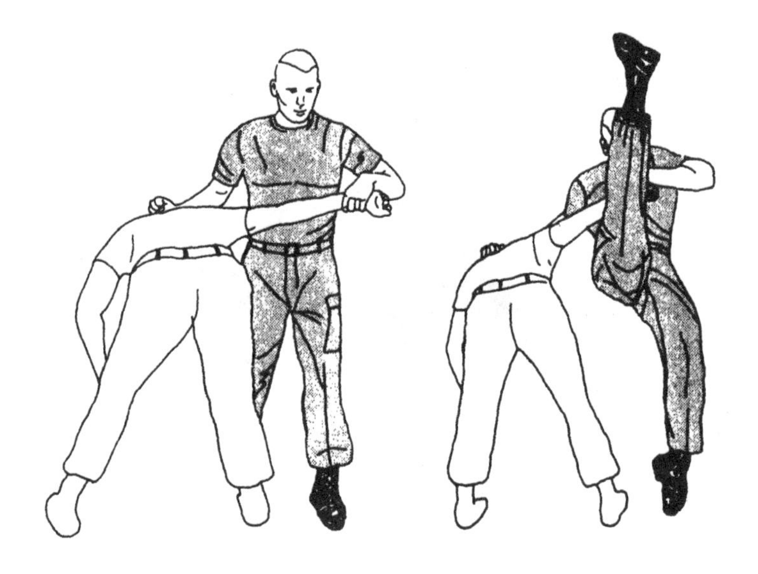

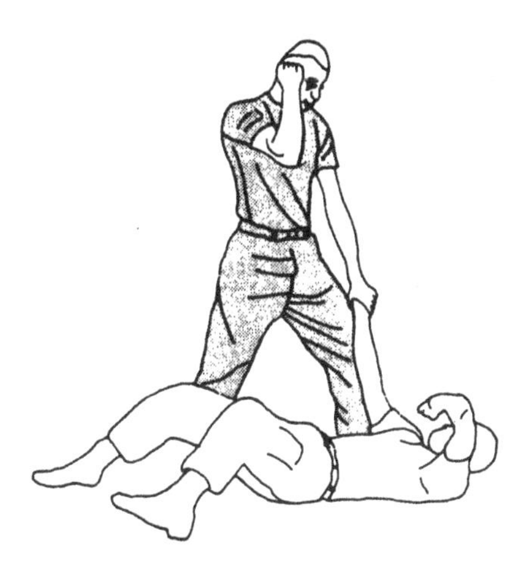

- Execute a heel stomp swiftly and violently to the opponent's head as a finishing technique.

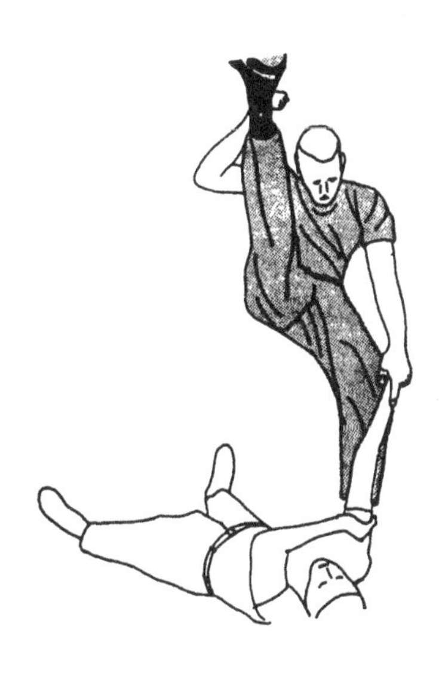

Defense for a Front Headlock

If the opponent attempts a front headlock, swiftly and forcefully deliver a right hand groin strike to loosen the hold. To defend against a front headlock—

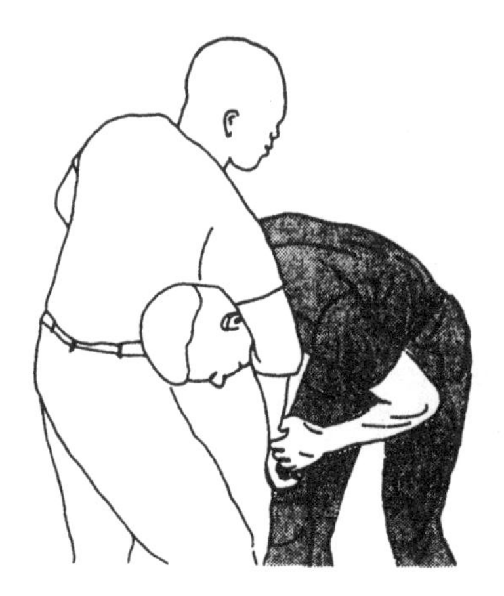

- Use the palm of your right hand to strike the opponent's groin.

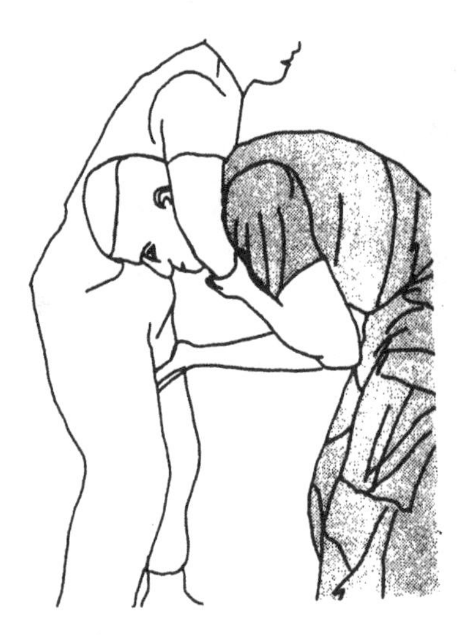

- Grab, squeeze, and twist the opponent's groin violently and forcefully.
- Grab the opponent's right wrist with your left hand.
- Step under the opponent's arm as the hold loosens.

- Deliver a powerful right forearm strike to the opponent's extended elbow. This damages the opponent's elbow and drives his upper body down.

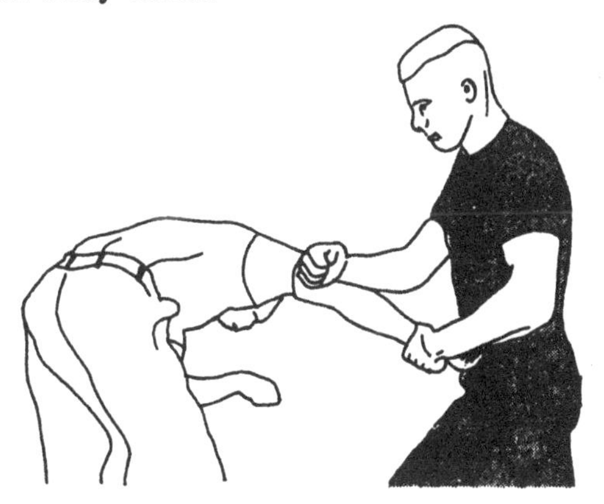

- Maintain control of the opponent's arm and apply pressure to his injured elbow.

- Execute a rear leg front kick to the opponent's face.

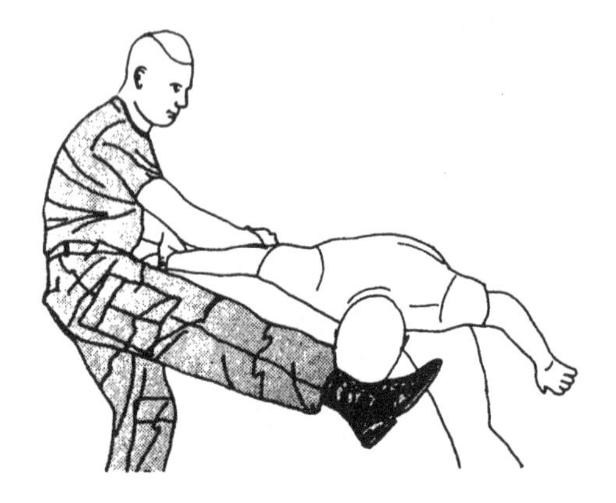

- Grab the opponent behind the neck, rotate your hips, and execute a leg sweep taking the opponent to the ground.

- Execute a heel stomp **swiftly and violently** to the opponent's head as a finishing technique.

Defense for a Rear Headlock

If the opponent applies a headlock from the rear, swiftly and forcefully gouge the opponent's eyes to loosen the hold. To defend against a rear headlock—

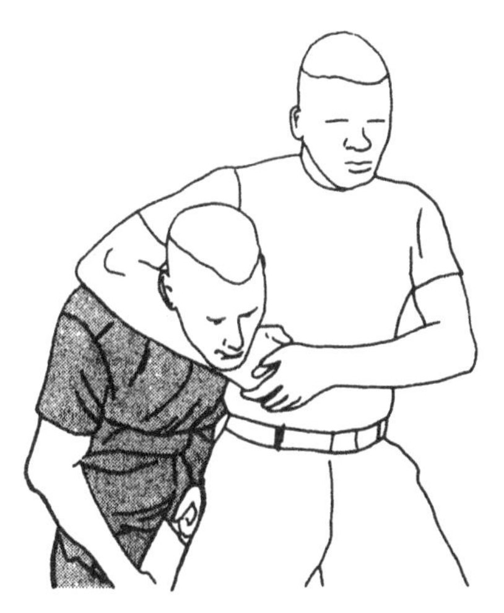

- Circle your arm that is closest to the opponent over his shoulder.
- Reach as far to the front of the opponent as possible to prepare for an eye gouge.

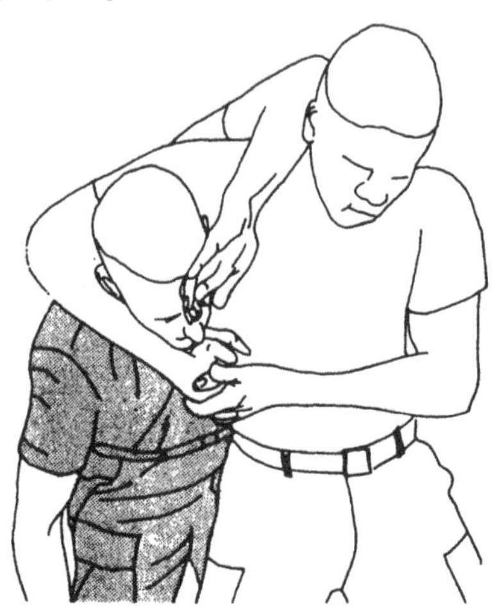

- Snap your arm forcefully back while gouging the opponent's eyes.
- Force your middle finger into the opponent's furthest eye socket.

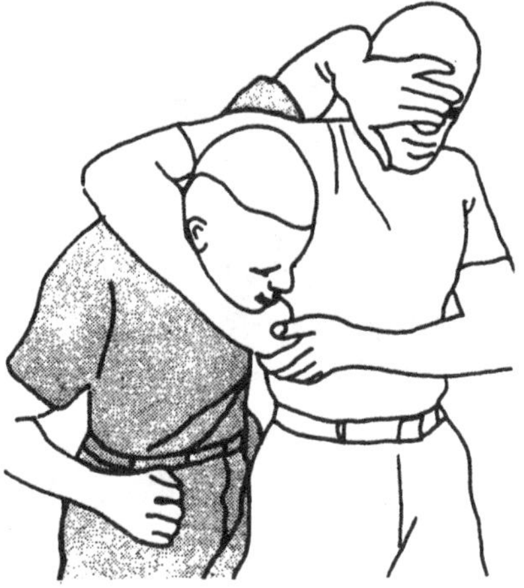

Note

Inserting your finger into the opponent's eye causes the opponent to release your head in order to remove your finger from his eye.

- Force the opponent's head back to expose his throat.

- Execute an open hand groin strike to bring the opponent's hands down and expose his throat.

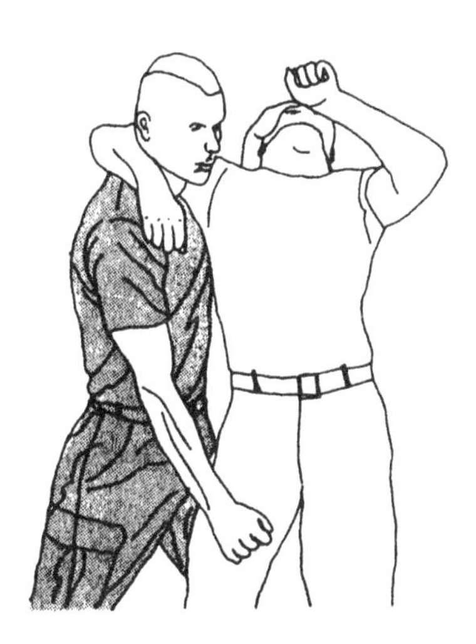

- Deliver a knifehand strike swiftly and violently to the opponent's throat as a finishing technique.

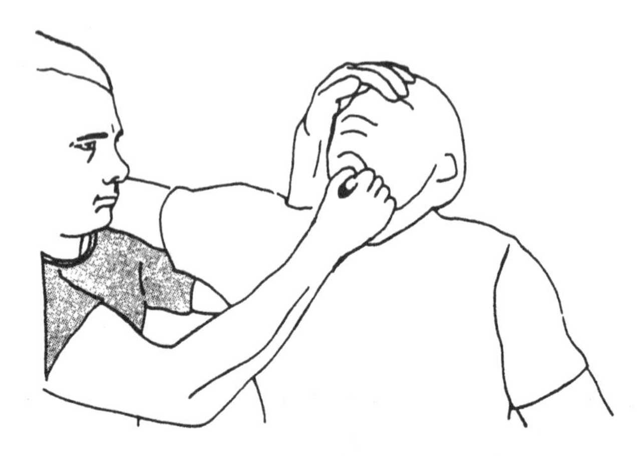

LINE II

Counters Against Punches and Kicks. LINE II teaches Marines to defend against attacks within the intermediate range (punching/kicking distances) of close combat and to destroy the opponent during the grappling stage.

Defense for a Lead Hand Punch

If the opponent attacks with a lead hand punch, parry with your rear hand to repel the attack. The key to this defense lies in your rapid response to the attack. To defend against a lead hand punch—

- Execute a rear hand parry while sliding forward with the lead foot.

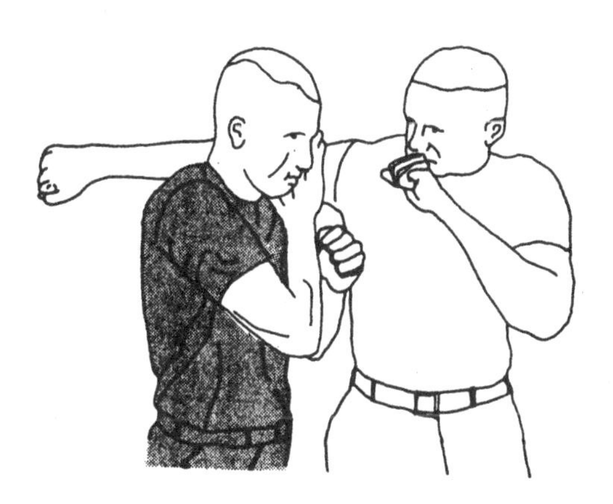

- Hook your left arm over the opponent's right shoulder while moving your right hand to the back of the opponent's neck to lock and control his right arm.

- Use both your arms to apply pressure and force the opponent's head down.

- Execute a knee strike to the face.

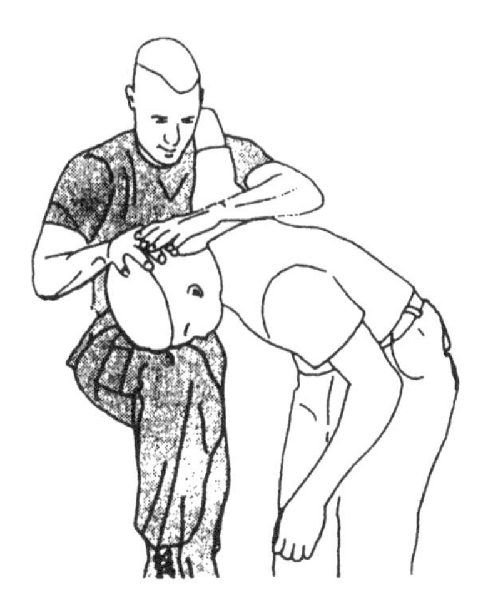

- Grab the opponent behind the neck, rotate your hips, and execute a leg sweep taking the opponent to the ground.

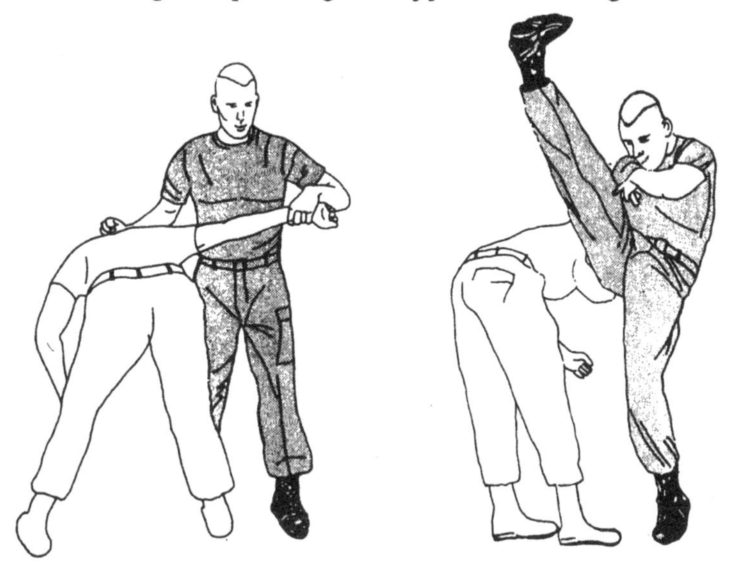

- Execute a heel stomp swiftly and violently to the opponent's head as a finishing technique.

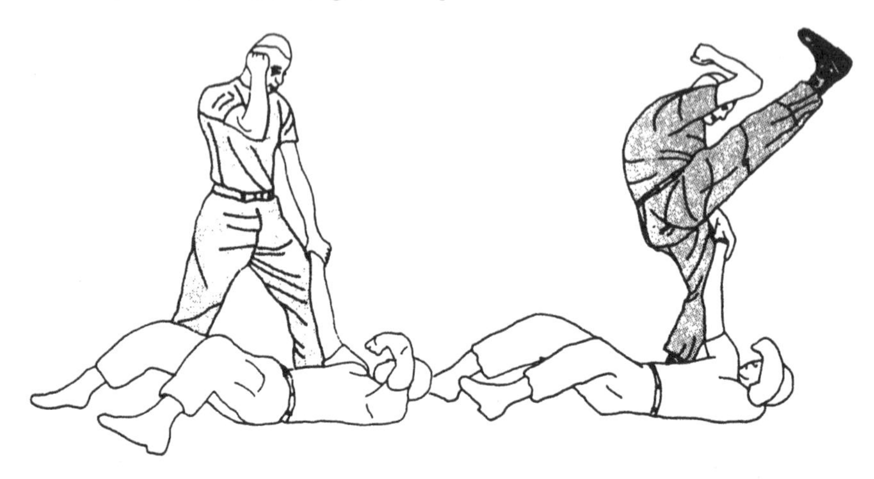

Defense for a Rear Hand Punch

If the opponent attacks with a rear hand punch, block with your lead hand to repel the attack. To defend against a rear hand punch—

- Execute an outside block with your lead hand.

- Step in with your rear foot and execute a forearm strike to the opponent's elbow with the inside of your rear forearm. This damages the opponent's elbow.

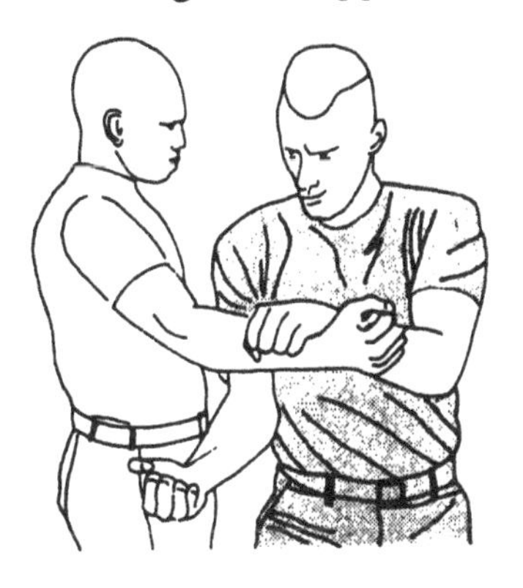

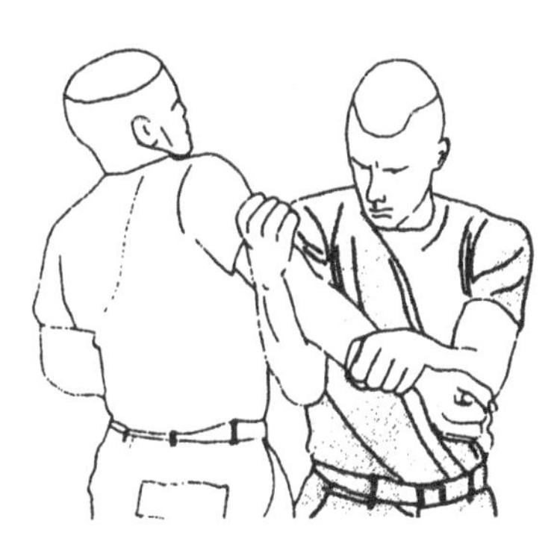

- Execute an elbow strike to the opponent's ribs.

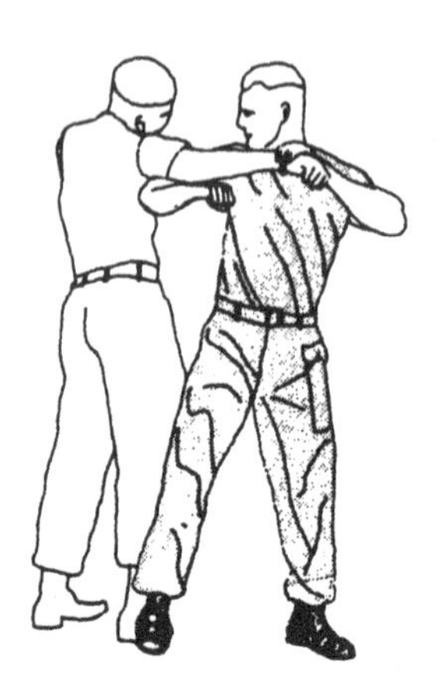

- Wrap your forearm and bicep around the opponent's upper arm.

- Rotate your hips and upper body and drive the opponent to the ground. To provide leverage for this movement, grip the opponent's injured upper arm, pull on his injured arm, and turn him over your hips and upper thigh.

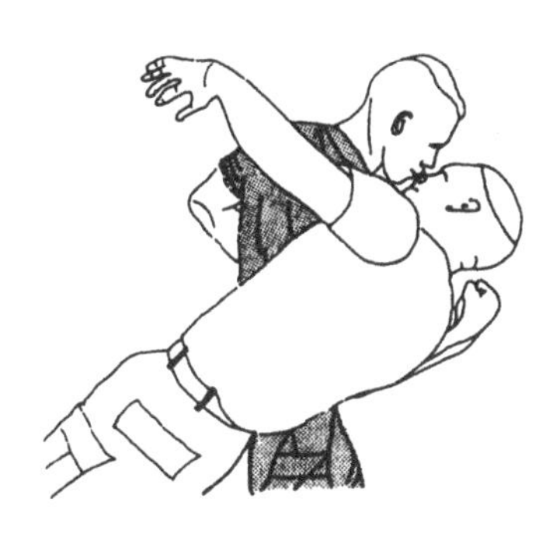

- Execute a heel stomp **swiftly and violently** to the opponent's head as a finishing technique.

Defense for an Uppercut Punch

If the opponent attacks with an uppercut punch, use your lead hand to block the attack. To defend against an uppercut punch –

- Execute a low block with your lead hand.

- Strike the inside of the opponent's elbow with the palm of your rear hand, not to cause damage but to create an opening between his arm and torso.

- Move your lead hand through the opening while your rear hand moves to the back of the opponent's neck to control his upper torso.

- Use both your arms to apply pressure and force the opponent's head down.
- Execute a knee strike to the opponent's face.

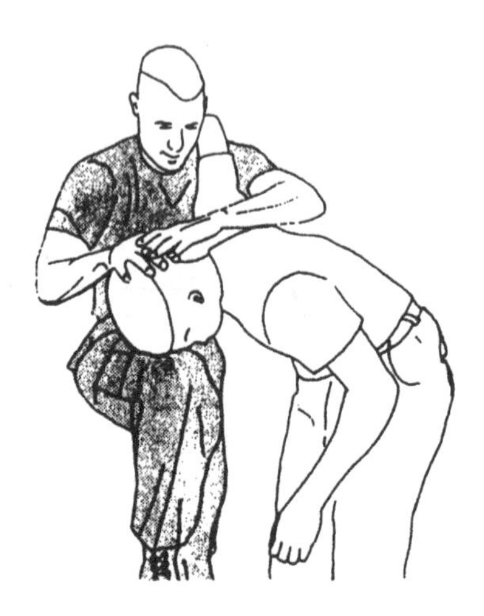

- Grab the opponent behind the neck, rotate your hips, and execute a leg sweep taking the opponent to the ground.

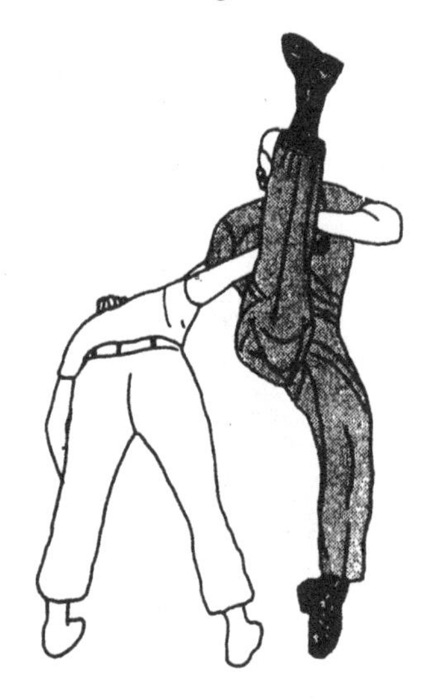

- Execute the heel stomp **swiftly and violently** to the opponent's head as a finishing technique.

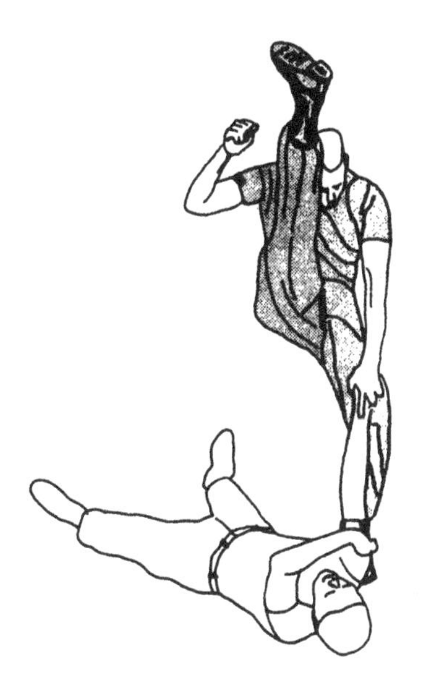

Defense for a Front Kick

If the opponent attacks with a front kick, parry with your lead hand to repel the attack. To defend against a front kick—

- Parry with the palm of your lead hand.

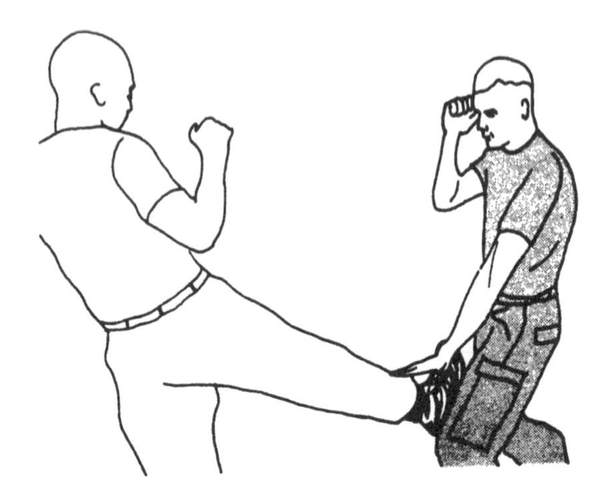

- Grab the back of the opponent's collar with your rear hand.
- Execute a rear leg side kick to the opponent's knee taking him to his knees.

- Execute an eye gouge with your lead hand.

- Force the head back exposing the opponent's throat.

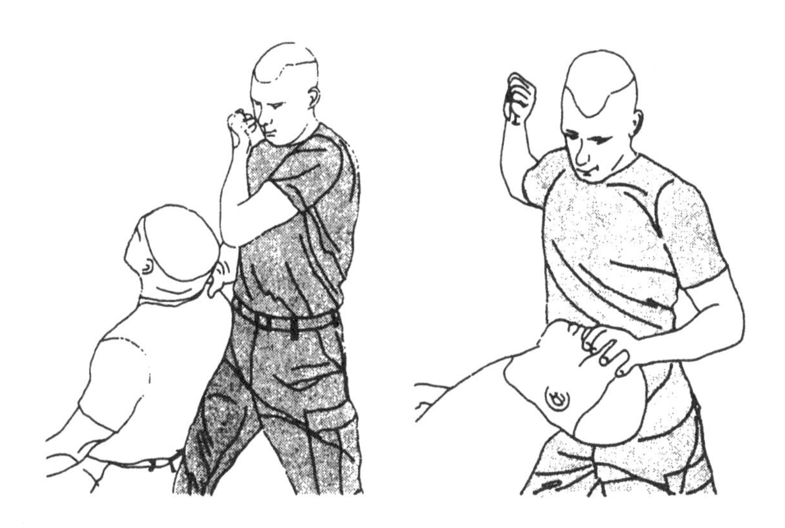

- Execute a knifehand strike to the opponent's throat as a finishing technique.

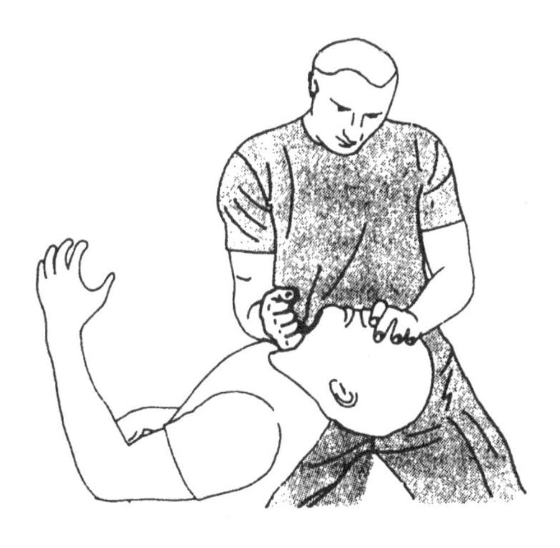

LINE III

Unarmed Defense Against a Knife. An opponent armed with a knife is a deadly adversary. The first step of unarmed defense against a knife is to neutralize the knife. The techniques used to neutralize a knife are **swiftly and violently** executed and cause permanent damage to the opponent's arm. Once the knife is neutralized, the opponent is destroyed using the techniques in LINE I and LINE II. You should expect to be cut during unarmed defense against a knife. There are five basic ways to attack with a knife: overhead attack, straight thrust, outside slash, inside slash, and uppercut. These basic attacks can be varied or combined. The following defensive techniques enable you to counter any type of knife attack.

Defense for an Overhead Attack

An opponent's overhead knife attack is a slashing or stabbing technique executed with an overhand motion toward the target. To defend against an overhead attack–

- Execute a high block with your lead hand.

- Step in with your rear leg.
- Execute a forearm strike with your rear arm to damage the opponent's elbow and neutralize the knife.

- Grasp the wrist of the opponent's injured arm.

- Execute an elbow strike to the opponent's ribs.

Note

An elbow strike to the ribs does not cause damage. It ensures that your arm is securely hooked around the opponent's upper arm.

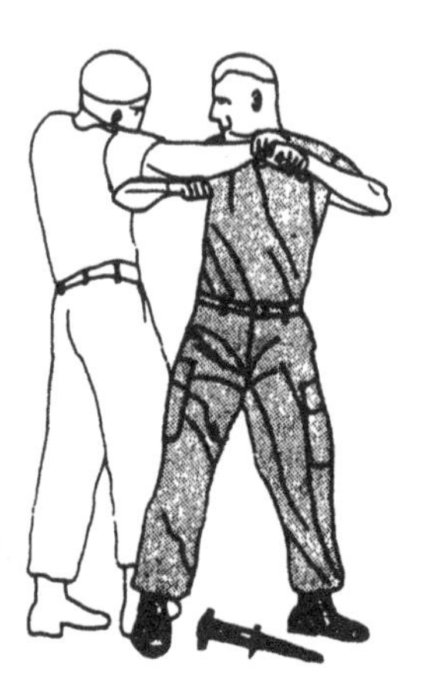

- Hook the opponent's upper arm with your forearm and bicep for leverage.

- Rotate your hips and upper body forcefully while pulling the opponent over your hip and upper thigh.

- Drive the opponent to the ground.

- Execute a heel stomp **swiftly and violently** to the opponent's head as a finishing technique.

Defense for a Straight Thrust

The straight thrust is the most dangerous offensive knife attack and is difficult to defend. To defend against a straight thrust—

- Execute a low block with your lead hand to deflect the opponent's thrust to the outside of your body.

- Step in with your rear leg.

- Execute a forearm strike with your rear forearm to damage the opponent's elbow and neutralize the knife.

- Grasp the wrist of the opponent's injured arm.

- Execute an elbow strike to the opponent's ribs.

Note

An elbow strike to the ribs does not cause damage. It ensures that your arm is securely hooked around the opponent's upper arm.

- Hook the opponent's upper arm with your forearm and bicep for leverage.

- Rotate your hips and upper body forcefully while pulling the opponent over your hip and upper thigh.

- Drive the opponent to the ground.

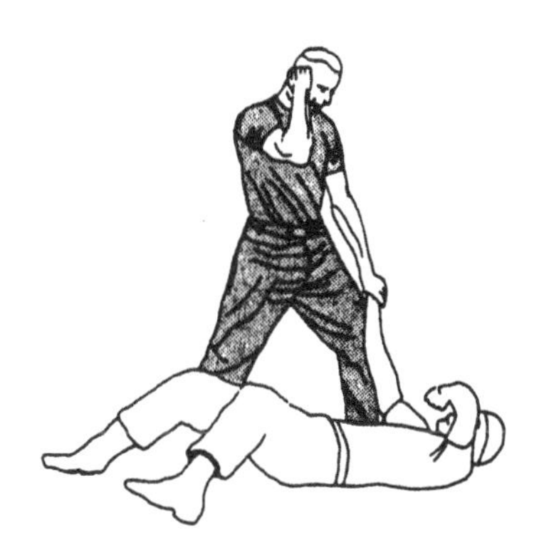

- Execute a heel stomp **swiftly and violently** to the opponent's head as a finishing technique.

Defense for an Outside Slash

An opponent's outside slash is an attack coming from the outside and is similar to a rearhand punch. To defend against an outside slash—

- Execute an outside block with your lead hand.

- Step in with your rear leg.
- Execute a forearm strike with your rear arm to damage the opponent's elbow and neutralize the knife.

- Grasp the wrist of the opponent's injured arm.

- Execute an elbow strike to the opponent's ribs.

Note

An elbow strike to the ribs does not cause damage. It ensures that your arm is securely hooked around the opponent's upper arm.

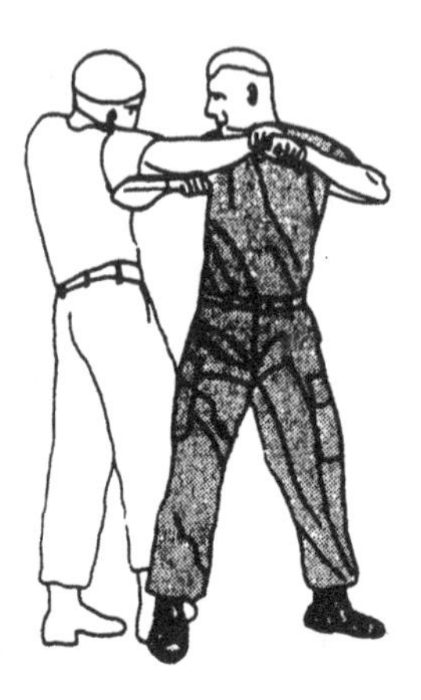

- Hook the opponent's upper arm with your forearm and bicep for leverage.

- Rotate your hips and upper body forcefully while pulling the opponent over your hip and upper thigh.

- Drive the opponent to the ground.

- Execute a heel stomp **swiftly and violently** to the opponent's head as a finishing technique.

Defense for an Inside Slash

An opponent's inside slash is a slashing movement that comes from the inside and is similar to a backhand slap. Typically, the inside slash is combined with the outside slash. To defend against an inside slash –

- Step in quickly and execute an outside block with your rear hand.

- Grasp the opponent's attacking wrist with your rear hand.
- Execute a forearm strike with your lead hand to damage the opponent's elbow and neutralize the knife.

- Apply pressure to the opponent's injured arm.
- Force the opponent's head down.
- Execute a rear leg front kick to the opponent's face.

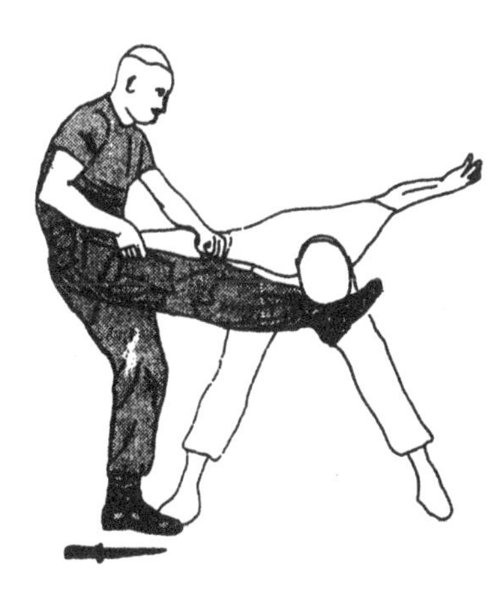

- Grab the opponent behind the neck.
- Rotate your hips and execute a leg sweep to take the opponent to the ground.

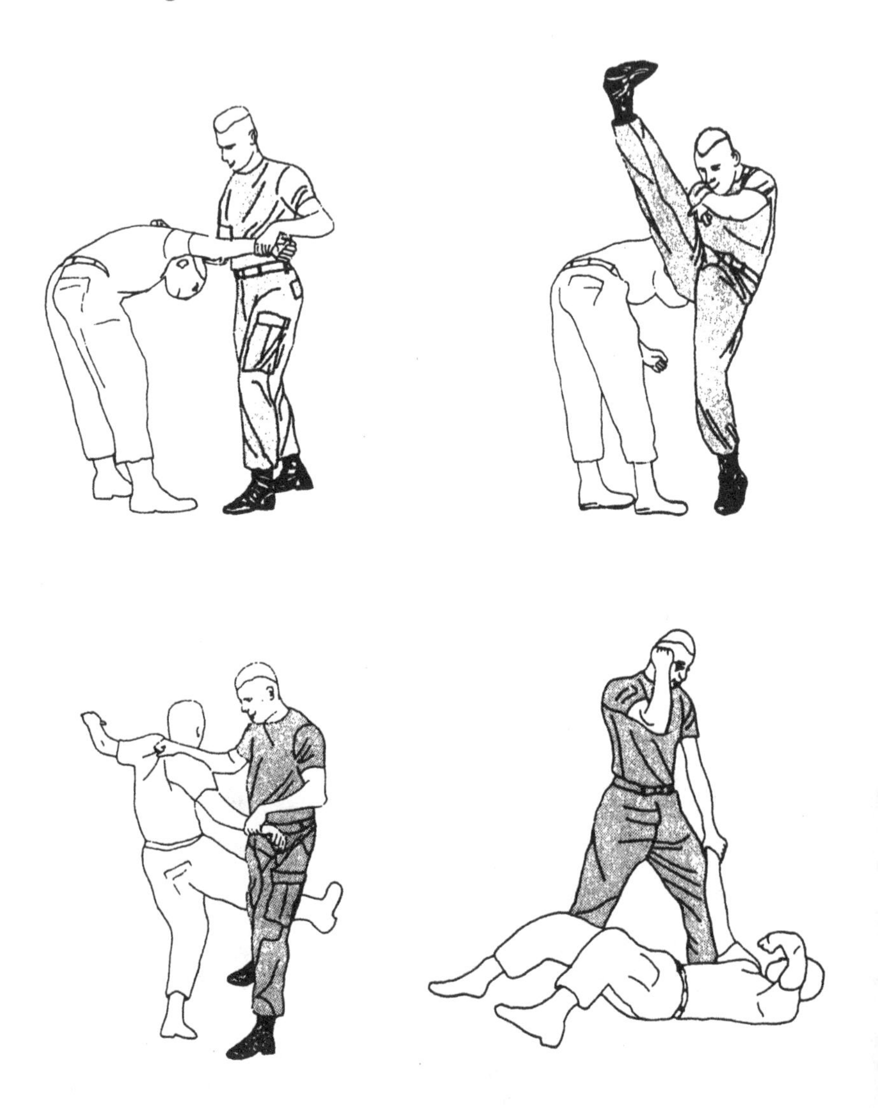

- Execute a heel stomp **swiftly and violently** to the opponent's head as a finishing technique.

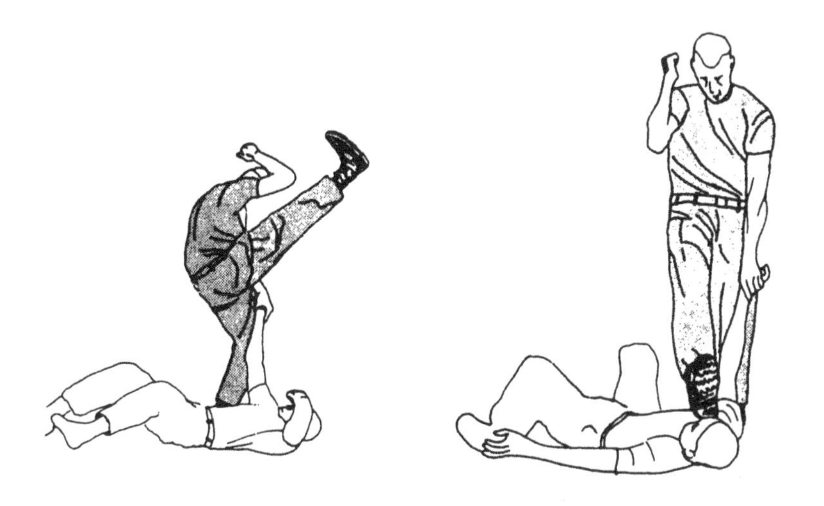

Defense for an Uppercut

The defense for an uppercut knife attack is the same as the defense for an uppercut punch (see page 80). The added length of the blade must be taken into account while blocking the attack. Typically, the uppercut is delivered closer than other knife attacks, so the defensive response must be quicker. To defend against an uppercut—

- Execute a low block with your lead hand.

- Step inside to the opponent's right.

- Strike the inside of the opponent's elbow with the palm of your rear hand. Strike hard enough to create an opening between the opponent's arm and torso.

- Move your lead hand through the opening while your rear hand moves to the back of the opponent's neck to control the opponent's upper torso.

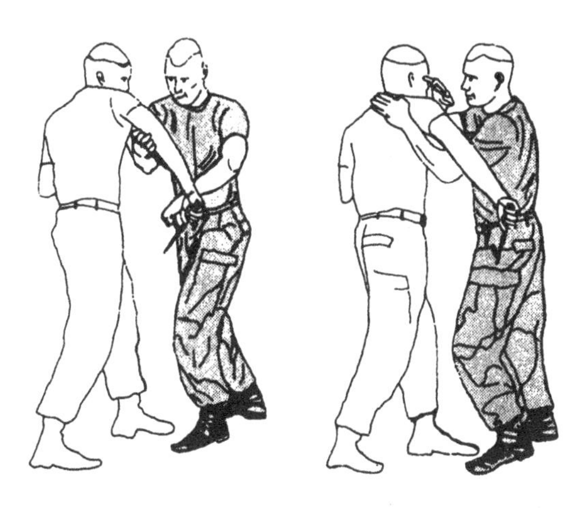

- Use both of your arms to apply pressure and force the opponent's head down.

- Execute a knee strike to the opponent's face.

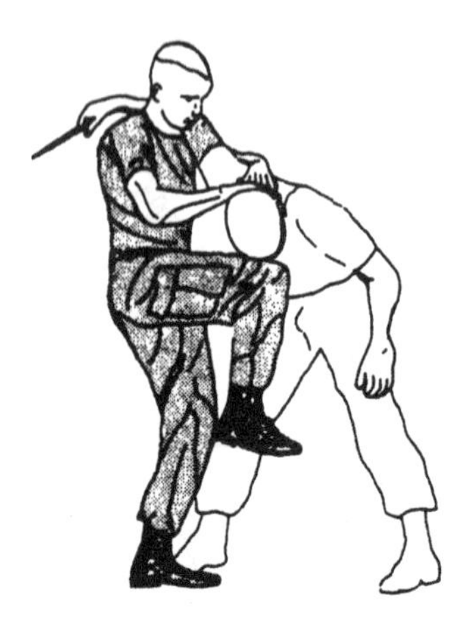

- Grab the opponent behind the neck.

- Rotate your hips and execute a leg sweep to take the opponent to the ground.

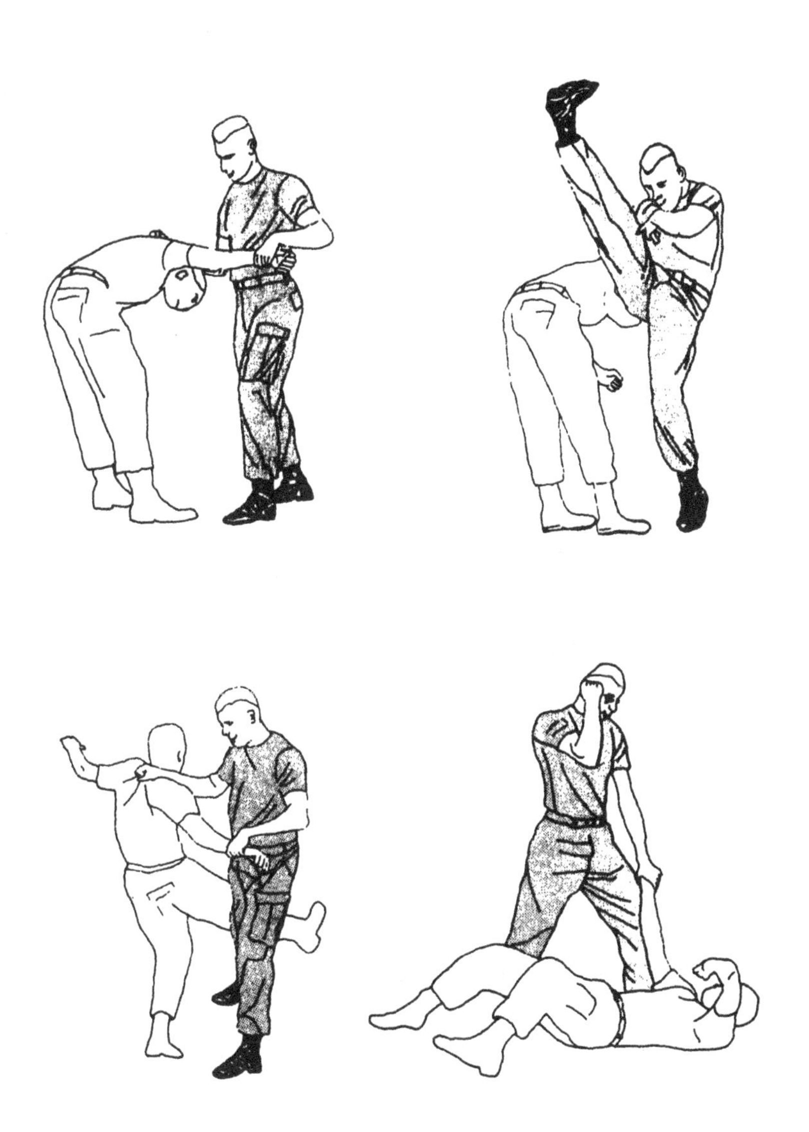

- Execute the heel stomp **swiftly and violently** to the opponent's head as a finishing technique.

LINE IV

Knife Fighting. A knife is an extension of the person. The same movements and techniques used in unarmed combat also apply to knife fighting. LINE IV techniques are designed to cause massive damage to the throat/neck area and quickly eliminate the opponent. During knife fighting, you must eliminate the opponent in the most ruthless and violent manner possible. For example, if properly executed, the initial rear hand slash severs the carotid artery and jugular vein and causes massive trauma to the trachea. The blade is retracted through the wounded area to increase the damage. The blade is then thrusted downward into the upper chest cavity through the opening caused by the first slash. This causes further damage to the trachea and punctures the lungs and aorta. The knife is plunged repeatedly into the opponent until he is destroyed.

Hammer Grip

The hammer grip is the most commonly used grip among nonskilled fighters. An advantage of this grip is the extended reach provided by the blade. A disadvantage of this grip is that the angle of the blade to the wrist does not afford maximum power for slashing. Another disadvantage is that it is hard to maintain if striking a hard object. To obtain the hammer grip –

- Grasp the handle of the knife. The blade points upward.
- Place thumb vertically below the base of the knife.

Icepick Grip

The icepick grip is the preferred grip for most close combat situations described in this manual. The icepick grip provides a strong grip, enables you to deliver powerful attacks, conceals the blade, and is difficult to defend. To obtain the icepick grip –

- Grasp the handle of the knife. The blade points downward.
- Ensure the cutting edge is forward.

Stance

The stance for knife fighting is the basic warrior stance. Your rear hand holds the knife. Your lead hand blocks and parrys while your rear hand delivers the decisive attack with the blade.

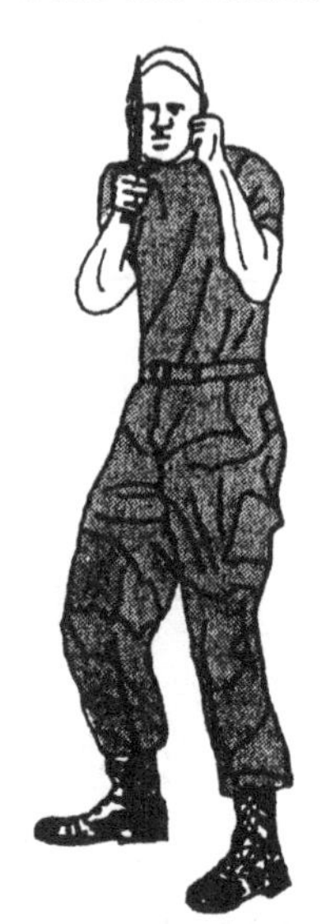

Defense for an Overhead Attack

To defend against an overhead attack—

- Execute a high block with the lead hand.

- Execute a rear hand slash (same motion as a rear hand punch, see p. 76) to the opponent's throat.

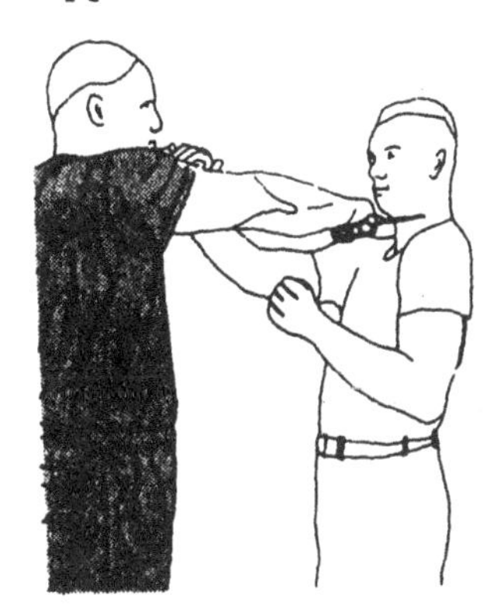

- Drive the blade through the opponent's throat/neck area.

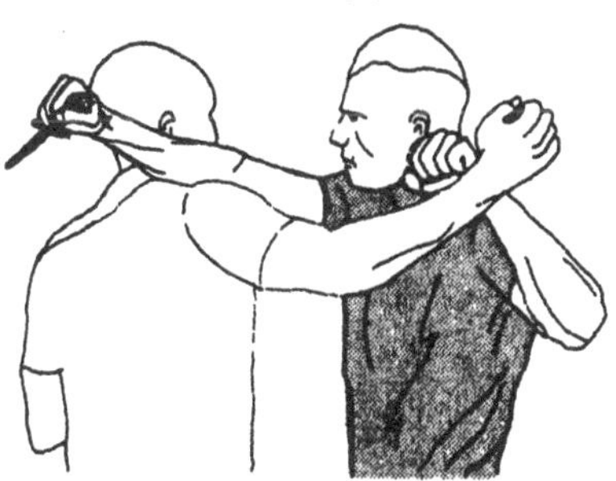

- Withdraw the blade through the opponent's wound to cause additional trauma to the damaged area.

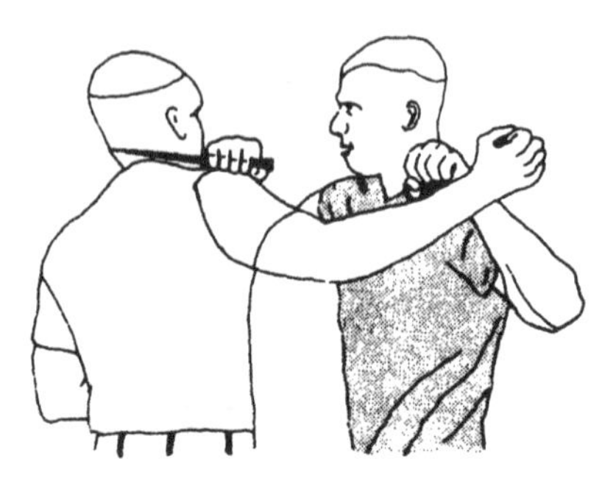

- Cock your arm back as the blade clears the opponent's body.

- Deliver a downward thrust into the opponent's upper chest cavity through the first wound as a finishing technique.

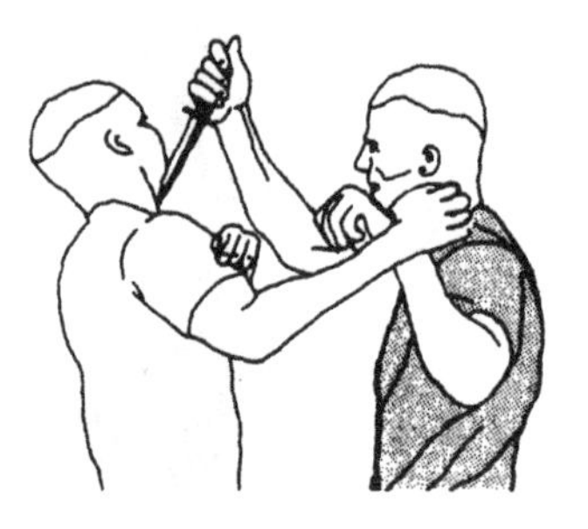

Defense for a Straight Thrust

To defend against a straight thrust—

- Execute a low block with the lead hand.

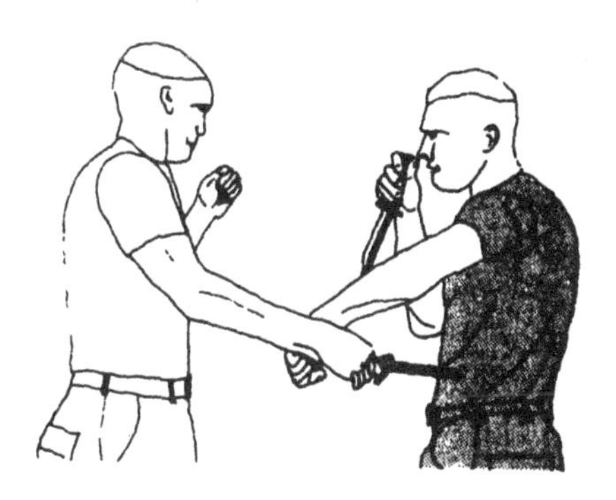

- Execute a rear hand slash (same motion as a rear hand punch, see p. 76) to the opponent's throat.

- Drive the blade through the opponent's throat/neck area.

- Withdraw the blade through the opponent's wound to cause more trauma.

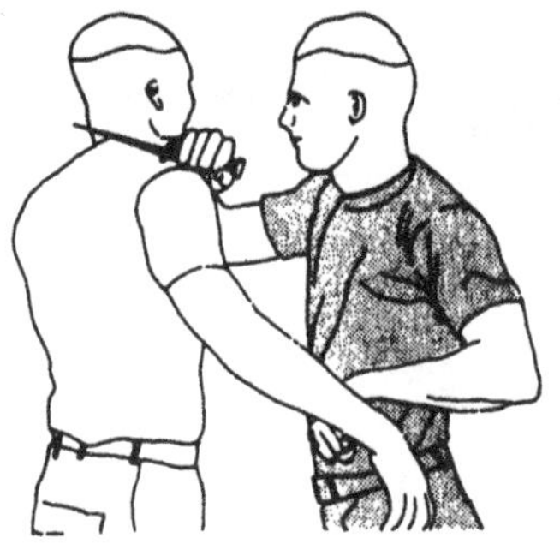

- Cock your arm back as the blade clears the opponent's body.

- Deliver a downward thrust into the opponent's upper chest cavity through the first wound as a finishing technique.

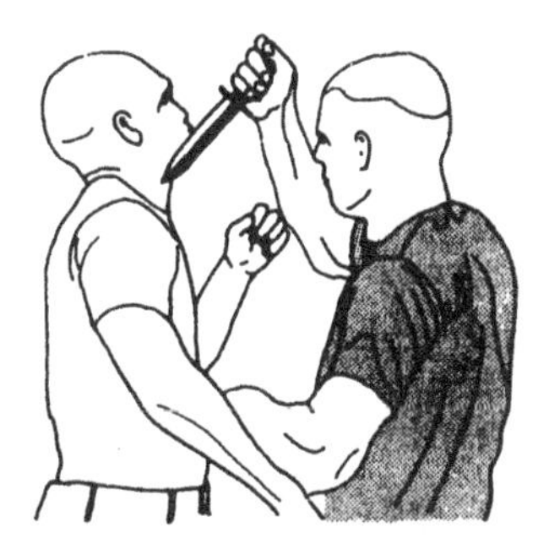

Defense for an Outside Slash

To defend against an outside slash—

- Execute an outside block with your lead hand.

- Execute a rear hand slash (same motion as a rear hand punch, see p. 76) to the opponent's throat.

- Drive the blade through the opponent's throat/neck area.

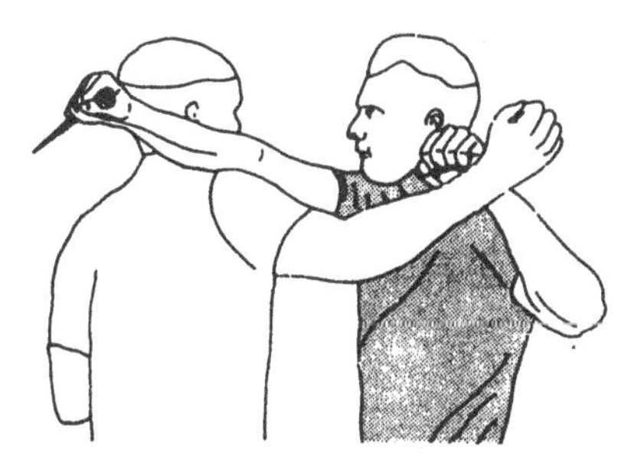

- Withdraw the blade through the opponent's wound to cause additional trauma.

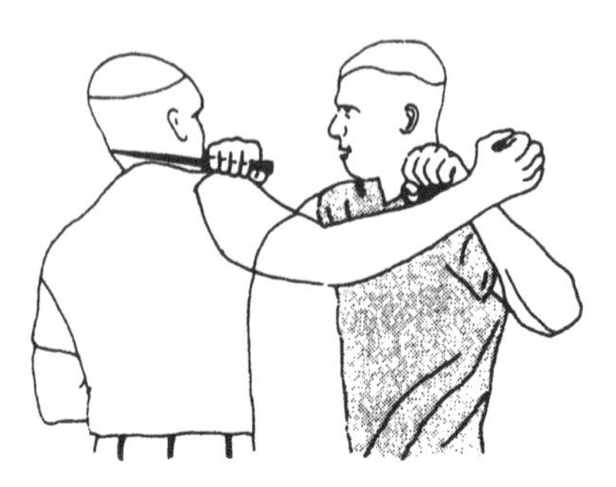

- Cock your arm back as the blade clears the opponent's body.

- Deliver a downward thrust into the opponent's upper chest cavity through the first wound as a finishing technique.

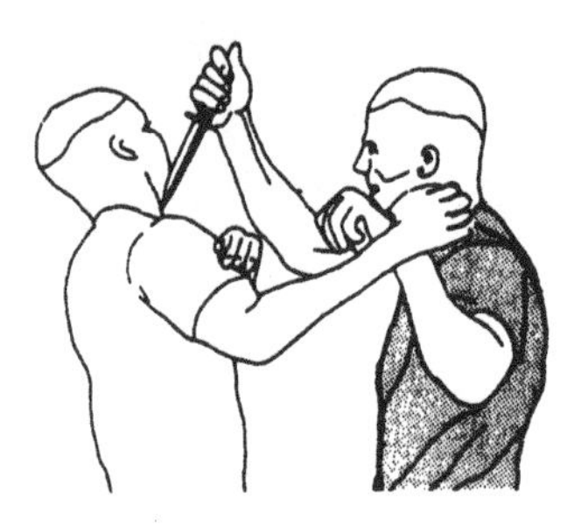

Defense for an Inside Slash

To defend against an inside slash–

- Execute a palm heel parry with your lead hand.

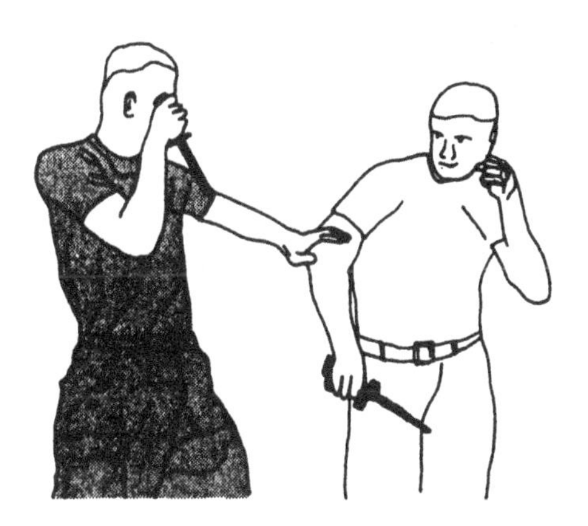

- Execute a rear hand slash (same motion as a rear hand punch, see p. 76) to the opponent's throat.

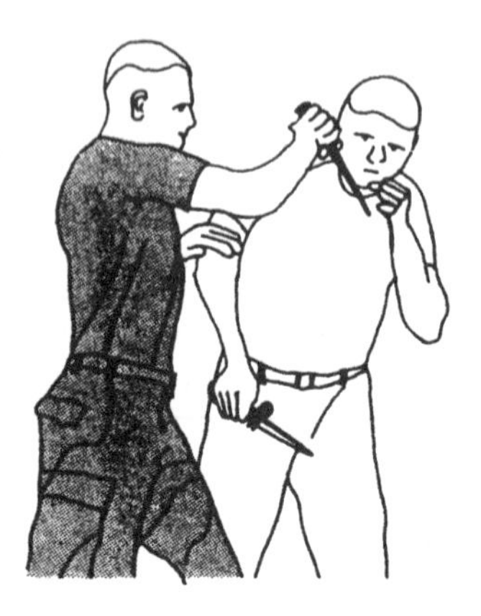

- Drive the blade through the opponent's throat/neck area.

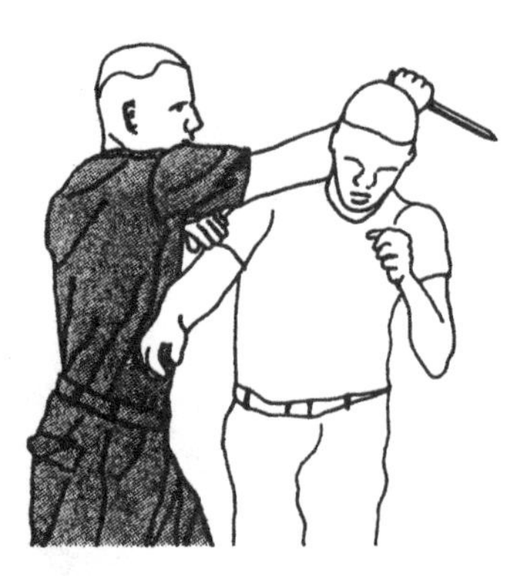

- Withdraw the blade through the opponent's wound to cause additional trauma.

- Cock your arm back as the blade clears the opponent's body.

- Deliver a downward thrust into the opponent's upper chest cavity and reenter the first wound as a finishing technique.

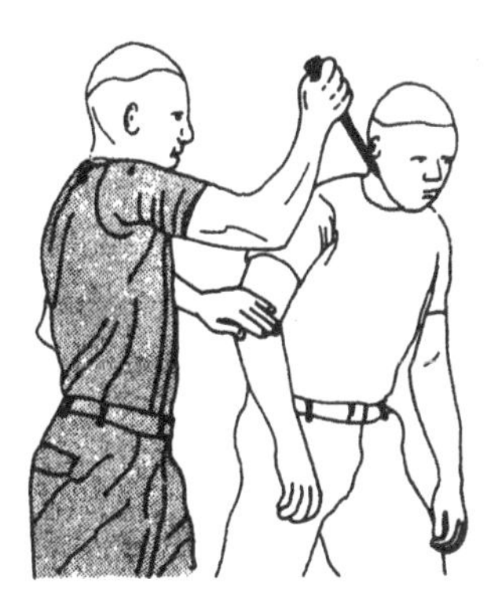

LINE V

Removal of Enemy Personnel. LINE V consists of a series of four techniques, two armed and two unarmed, designed to remove enemy personnel as quickly and quietly as possible. Although LINE V describes the basic techniques used to remove enemy personnel, silent removal of personnel should be conducted by specially-trained individuals.

Unarmed Removal from the Rear

To execute unarmed removal from the rear—

- Stalk the opponent from the rear.
- Remain slightly to the right of the opponent.
- Keep your body below the opponent's line of sight.
- Modify the basic warrior stance by crouching.
- Remain alert to the opponent's movements.

- Execute an eye gouge with your lead hand while forcing the opponent's head back to expose his throat.

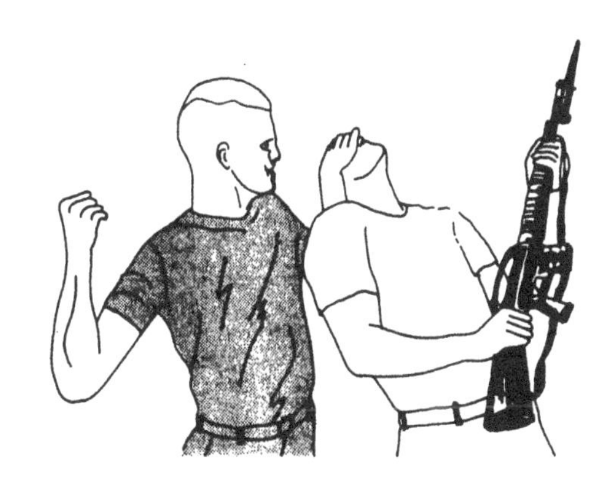

- Deliver several knifehand strikes to the opponent's throat with your rear hand.

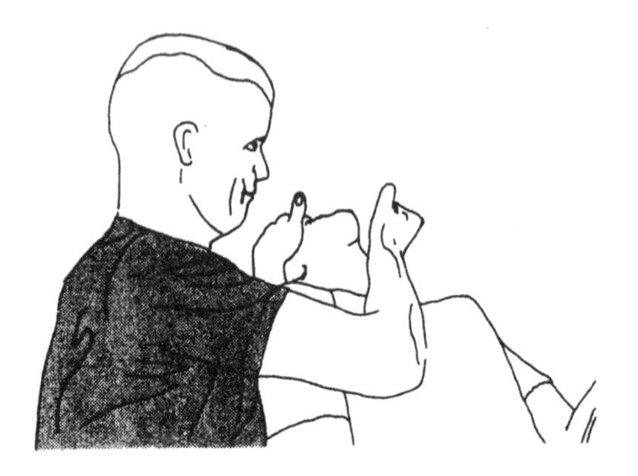

- Pull the opponent to the ground.
- Continue to deliver knifehand strikes to the opponent's throat until he is dead.
- Use your upper body to cover the opponent's upper body and head. This prevents any postmortem movement and reduces the sound of expelling air.

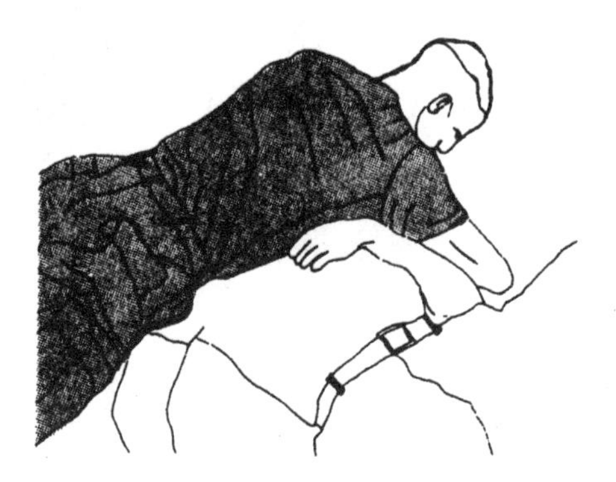

Unarmed Removal from the Prone

To execute unarmed removal from the prone—

- Assume the prone position and face the approaching opponent.
- Push upward with your left arm and right knee as the opponent approaches.
- Execute an open hand groin strike with your right hand.

- Place your left hand on top of the opponent's head and your right hand under his chin.
- Use a violent snapping motion to pull the opponent's chin toward you while pushing the back of his head away.

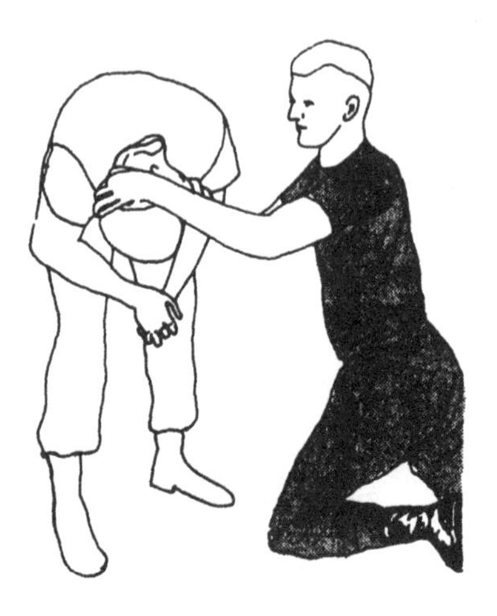

- Take the opponent to the ground.
- Execute an eye gouge with your left hand while forcing the opponent's head back.

- Execute knifehand strikes to the opponent's throat with your right hand as a finishing technique.

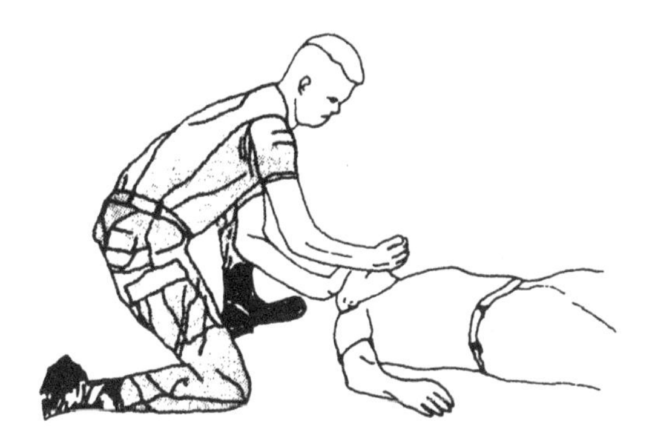

- Use your upper body to cover the opponent's upper body and head. This prevents any postmortem movement and reduces the sound of expelling air.

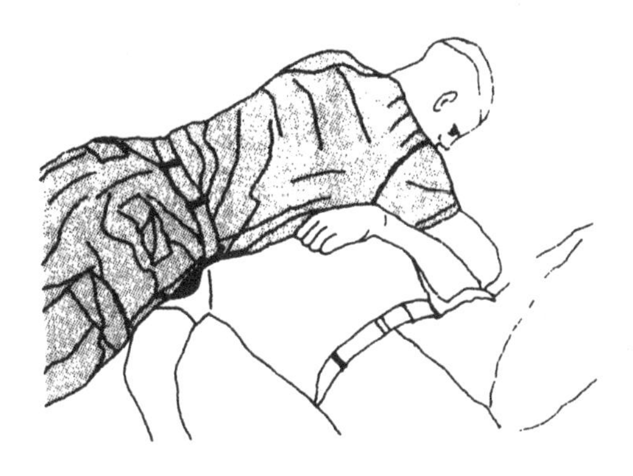

Armed Removal from the Rear

To execute armed removal from the rear—

- Stalk the opponent from the rear.

- Keep your body below the opponents line of sight.
- Remain alert to the opponents movements.
- Hold the knife in an icepick grip with the cutting edge facing the forearm.

- Execute an eye gouge with your lead hand.
- Gouge the opponent's right eye socket and snap his head back to expose the throat.

- Plunge the blade into the left side of the opponent's throat.

- Snap the opponent's head violently to the left.
- Rip the blade across the opponent's throat to the right.

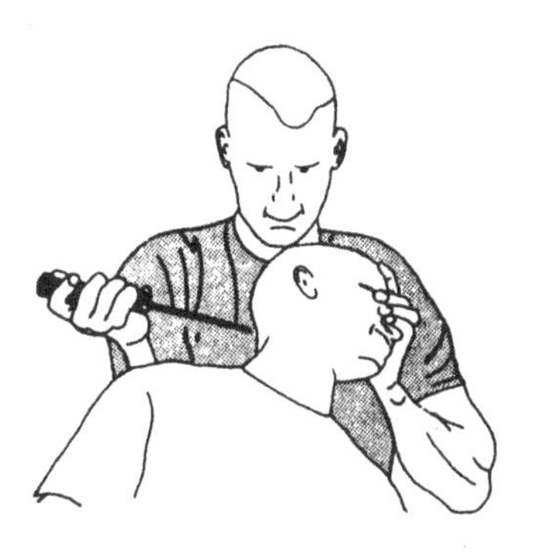

- Execute the finishing technique by raising the blade over the opponent's head and thrusting it into his upper chest cavity through the wound caused by the throat cut. This will puncture the opponent's lungs and aorta.

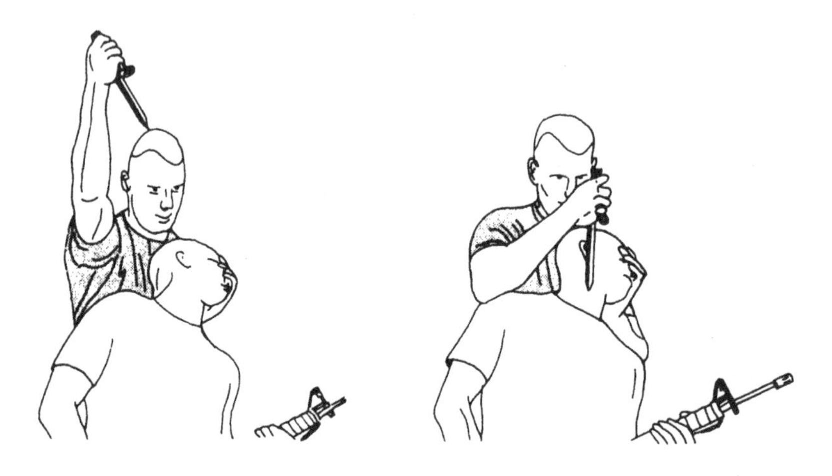

- Follow the opponent to the ground.

- Use your upper body to cover the opponent's upper body and head. This prevents postmortem movement and reduces the sound of expelling air.

Armed Removal from the Prone

To execute armed removal from the prone–

- Assume the prone position and face the approaching opponent.
- Hold the knife in your rear hand using the hammer grip. The cutting edge is up and toward the thumb.

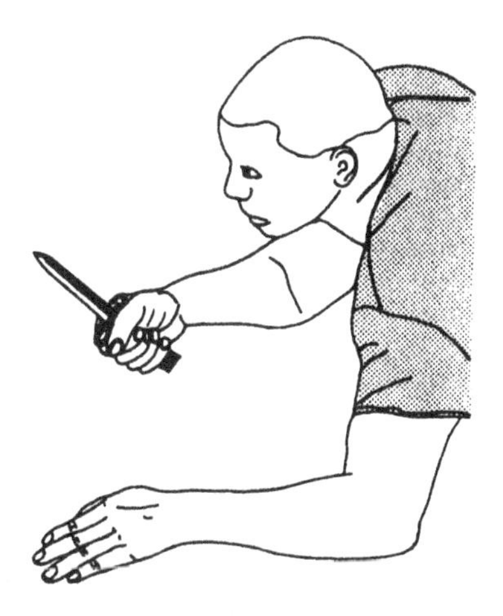

- Push upward with your lead arm and right knee as the opponent approaches.

- Thrust the blade into the opponent's groin area between the groin and anus.

- Rip the blade toward the opponent's groin as he bends at the waist.

- Grab the back of the opponent's head with your lead hand.

- Thrust the blade into the far side of the opponent's throat.

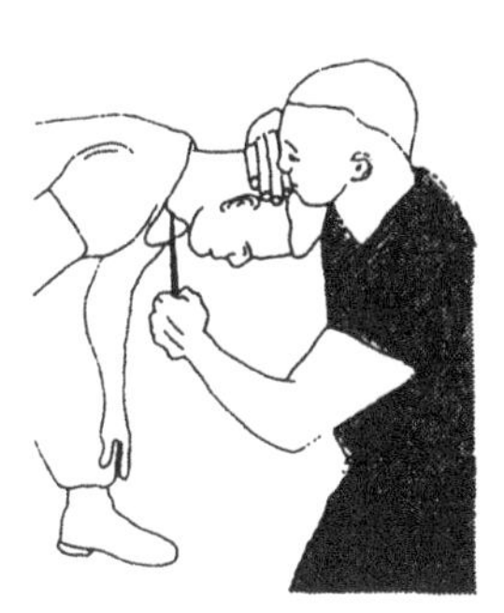

- Push the opponent's head away while pulling the blade through his throat.

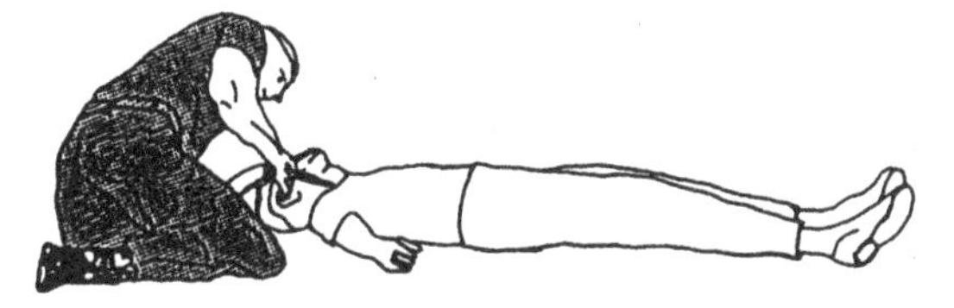

- Use your upper body to cover the opponent's upper body and head. This prevents postmortem movement and reduces the sound of expelling air.

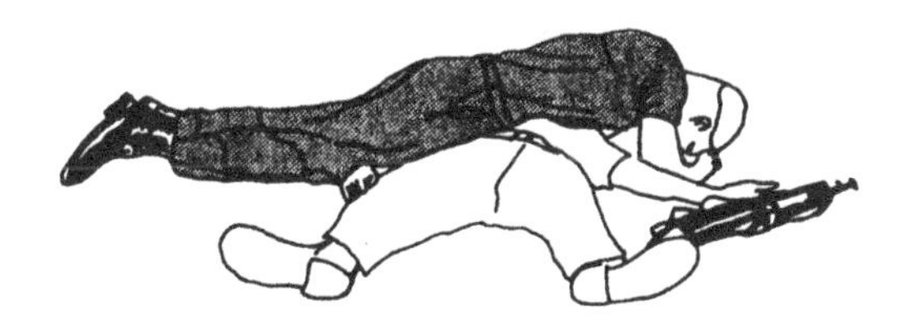

BAYONET FIGHTING

The bayonet still has a place in the modern battle's arsenal of weapons. Training with a bayonet instills confidence in an individual. This confidence allows an individual to close with and destroy the enemy under a variety of conditions. Through proper training, Marines develop the courage and confidence required to effectively use a bayonet to protect themselves and destroy the enemy. In situations where friendly and enemy troops are closely mingled and rifle fire and grenades are impractical, the bayonet becomes the weapon of choice.

To be successful with the bayonet, you must be aggressive, ruthless, savage, and vicious. You must follow each vicious attack with another vicious attack, remembering that if you do not kill your opponent, your opponent will kill you. Hesitation, delay, and excessive maneuvering can result in your death. Your primary aim is to sink your blade into a vital area of the enemy (the throat is the best target). Strikes with the rifle butt or slashes from the blade can cause the enemy to waiver in his protective posture. Once the enemy waivers, you can use the blade to attack a vital area. Remember, the rifle and bayonet provide you with a good shield, and a way to block and parry attacks by the enemy. To be successful, you should strike the first blow and follow up with the kill. The best defense is not to allow the opponent to take the offense.

The Guard Position

The guard position is an armed version of the basic warrior stance. All movements originate from the guard position. Hold the weapon in approximately the same position as port arms except that the sling and bayonet's cutting edge face the enemy and the weapon is held further away from the body in order to absorb the shock of an attack.

Training and practice are the only ways to attain proper form, accuracy, agility, and speed between the rifle and feet. Practice and training increases coordination, balance, speed, and endurance. Moves are practiced until they become instinctive.

Perform the following steps to attain the guard position:

- Grasp the handguard under the upper sling swivel with your lead hand.

- Grasp (approximately 2 inches from the charging handle) the small of the stock with your rear hand. This helps prevent finger injuries when striking.

- Hold the rifle away from your body approximately 10-15 inches. The butt is in front of your right hip. The muzzle bisects the angle formed between your head and left shoulder.

- Bend and relax your arms for rapid movement.

- Place your feet shoulder-width apart with the toe of your right foot in line with the heel of your left foot.

- Bend the knees slightly to evenly distribute your body's weight on both legs.

- Hold the upper body erect.

You must be able to assume the guard position instinctively and move in all directions while maintaining the guard position. During movement, your legs and feet should not be crossed and your upper body should remain in the guard position.

Forward Movement (Advance)

Slide forward approximately 12-15 inches with the lead foot. As soon as your lead foot is in place, quickly move your rear foot forward to return to the guard position.

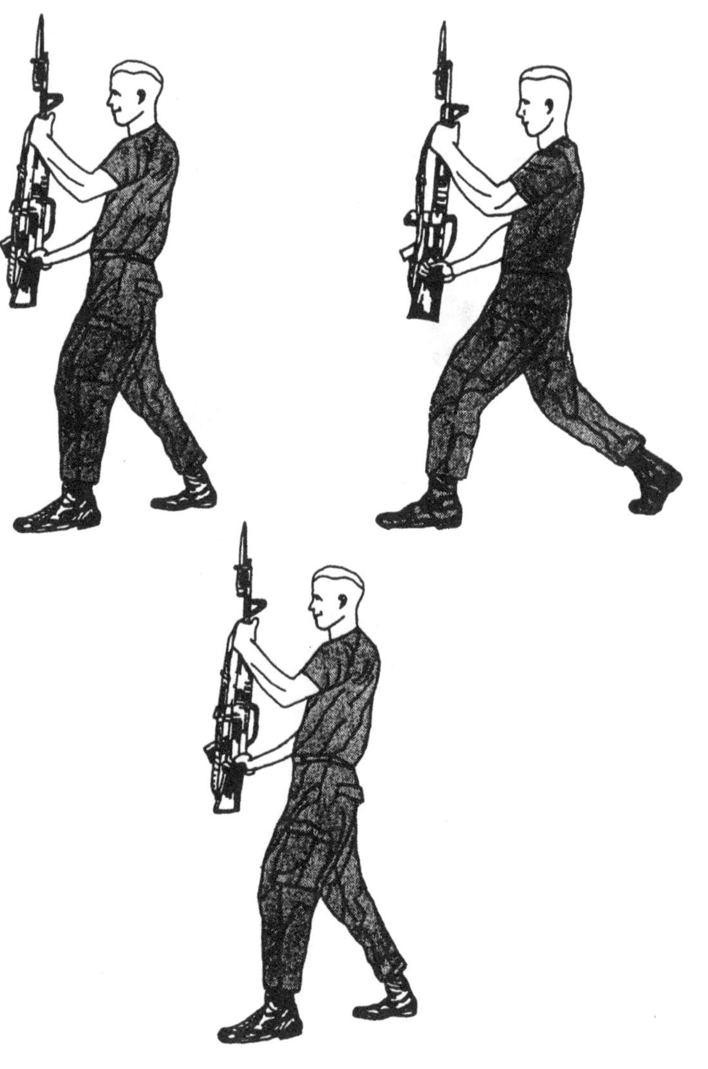

Right/Left Movement (Side Step)

Slide to the side approximately 12-15 inches with your foot. As soon as your foot is in place, quickly follow with your other foot to return to the guard position. The side step is best used in combination with a parry or offensive strike.

Turning Movement (Whirl)

Sometimes you must change direction in order to face the opponent. By using the whirl, you can turn to either the right, left, or rear. To execute the whirl, quickly pick up your foot opposite from the direction desired and pivot in the desired direction on the ball of your other foot. As soon as you face the desired direction, return to the guard position.

Offensive Skills

There are five basic attacks used in bayonet fighting: the slash, the straight thrust, the horizontal butt stroke, the vertical butt stroke, and the smash. These attacks can be used and should be practiced in combination with each other and in conjunction with individual offensive and defensive movements. To ensure success, attacks are swiftly and relentlessly delivered to the opponent's target areas until he is destroyed.

The Slash

The slash is created by a quick, slicing motion of the bayonet and relies on speed rather than force. It is not a chopping motion. The main target area of the slash is the opponent's neck. You can use the slash to kill the opponent or to create an opening in his defense. To execute the slash—

Note

The slash can be executed in conjunction with the advance, side step, or whirl.

- Extend your lead hand forward while pulling the rifle stock under your rear arm with your rear hand.
- Retract the bayonet by reversing the movement.
- Return to the guard position or follow up with another attack.

The Straight Thrust

The straight thrust is the most difficult technique to defend. If delivered correctly, it can effectively disable and kill the opponent. Since the opponent's throat, groin, and face are typically unprotected, they are your best targets. The opponent's chest and stomach are also excellent targets, but they may be protected by body armor or combat equipment. To execute the straight thrust—

Note

The straight thrust can be executed in conjunction with the advance, side step, or whirl.

- Rotate your upper body so your lead shoulder rotates forward.
- Lower the rifle until the bayonet is parallel to the ground and pointing at your opponent.
- Use your arms, shoulders, and hips to generate power and speed while thrusting the bayonet forward and into your target.
- Rotate the rifle to twist the blade in the wound and drive the bayonet deeper into the opponent until he collapses.
- Retract the bayonet by returning your arms to the guard position.

The Horizontal Butt Stroke

You can use the horizontal butt stroke to weaken enemy defenses, injure the enemy, or set the enemy up for the killing blow. The main targets of the horizontal butt stroke are the opponent's head and neck. The horizontal butt stroke is an excellent technique when used in conjunction with the slash or the straight thrust. If the opponent deflects a slash or straight thrust, the momentum of the attack sets up the horizontal butt stroke. To execute the horizontal butt stroke—

Note

The horizontal butt stroke can be executed in conjunction with the advance, step, or whirl.

- Swing the rifle forward horizontally with your rear hand while pulling the rifle over your left shoulder with your lead hand and rotating your shoulder and hip. This generates power and speed.

Note

Do not step forward.

- Strike the opponent with the toe of the stock.

- Retract the rifle immediately.

- Return to the guard position or follow up with another attack.

The Vertical Butt Stroke

You can use the vertical butt stroke to weaken enemy defenses, injure the enemy, or set the enemy up for the killing blow. The main target areas of the vertical butt stroke are the opponent's groin and face. The vertical butt stroke is an excellent technique when used in conjunction with the slash. To execute the vertical butt stroke—

Note

The vertical butt stroke can be executed in conjunction with the advance, step, or whirl.

- Push the rifle forward and upward with your rear hand while pulling the rifle over your left shoulder with your lead hand and rotating your shoulder and hip. This generates power and speed.

Note

Do not step forward.

- Strike the opponent with the toe of the stock.
- Retract the rifle immediately.
- Return to the guard position or follow up with another attack.

The Smash

The smash is a follow-up technique to the horizontal or vertical butt stroke. After delivery of the horizontal or vertical butt stroke, the rifle is cocked with the butt pointing toward the opponent, ready for the smash. The main target area of the smash is the head. To execute the smash—

- Draw the rifle back over your left shoulder.
- Drive the rifle butt into the opponent's face.
- Return to the guard position or follow up with another attack.

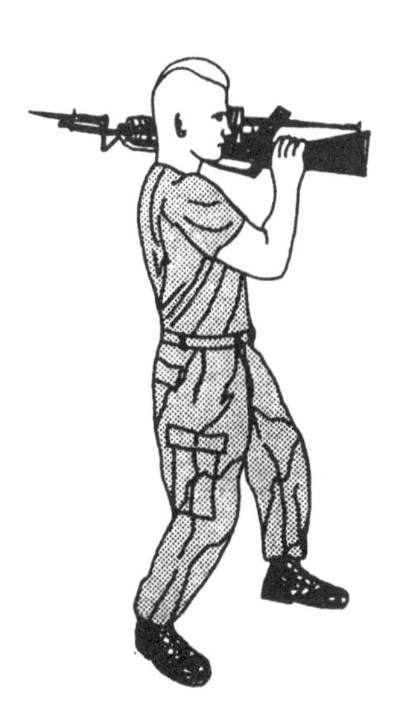

Defensive Skills

Defensive movements with the bayonet protect you and allow you to regain the initiative. There are four basic defensive moves: high block, low block, left parry, and right parry.

Blocking is effective against the slash and vertical butt stroke. Parrying is effective against the straight thrust, smash, or horizontal butt stroke.

Defensive moves are executed with as much speed and force as possible. While executing a defensive move, do not overextend. Overextending creates openings for the opponent. Only extend enough to neutralize the attack. Practice defensive moves from a stationary guard position and while moving.

High Block

The high block counters overhead or high attacks (e.g., the slash). To execute the high block –

- Snap the rifle upward forcefully. The rifle is parallel to the ground and clears the top of your head.
- Extend your arms up and out at approximately a 45-degree angle to your body. Your upper body is erect.
- Apply tension to your elbows and shoulders. Do not lock your elbows.

After you block the opponent's attack, you can counterattack with the slash and horizontal butt stroke to regain the initiative and destroy the opponent.

Low Block

The low block counters low attacks (e.g., vertical butt stroke). To execute the low block –

- Snap the rifle downward forcefully. The rifle is parallel to the ground and below your waist.
- Extend your arms down and out at approximately a 30 to 45 degree angle to your body. Your upper body is erect.
- Apply tension to your elbows and shoulders. Do not lock your elbows.

After you block the opponent's attack, you can counterattack with the slash and horizontal butt stroke to regain the initiative and destroy the opponent.

Left Parry

The left parry defends against incoming attacks (e.g., straight thrust, horizontal butt stroke, smash) from the left of the weapon. To execute the left parry—

- Snap the rifle forward and to your left forcefully while rotating your shoulders and hips to generate speed and power. The rifle is perpendicular to the ground and clears the left side of your body. Your upper body is erect.
- Extend your rear arm without locking your elbows.
- Cock your lead arm for a counterattack.

After you deflect the opponent's attack, you can counterattack with the slash and the horizontal or vertical butt stroke to regain the initiative and destroy the opponent.

Right Parry

The right parry defends against incoming attacks (e.g., straight thrust, horizontal butt stroke, and smash) coming from the right of the weapon. To execute the right parry—

- Snap the rifle forward and to your right forcefully while rotating your shoulders and hips to generate speed and power. The rifle is perpendicular to the ground and clears the right side of your body. Your upper body is erect.
- Extend your lead arm without locking your elbows.
- Bend your rear arm slightly for a counterattack.

After you deflect the opponent's attack, you can counterattack with the straight thrust and the horizontal or vertical butt stroke to regain the initiative and destroy the opponent.

Combination Movements

Bayonet movements can be combined in order to maintain the initiative and destroy the opponent. These movements are not memorized but practiced until they become instinctive. The key to a successful combination is the aggressiveness of the bayonet fighter. The following combinations are examples of effectively transitioning from one movement to another in order to destroy the opponent.

Combination 1
Guard Position
Slash
Vertical or Horizontal Butt Stroke
Slash
Straight Thrust
Recover to Guard Position

Combination 2
Guard Position
Straight Thrust
Vertical Butt Stroke
Smash
Slash
Straight Thrust
Recover to Guard Position

Combination 3
Guard Position
High Block a Slash Attack
Slash
Vertical or Horizontal Butt Stroke
Smash
Slash
Straight Thrust
Recover to Guard Position

Combination 4
Guard Position
Low Block a Vertical Butt Stroke
Straight Thrust or Slash
Horizontal or Vertical Butt Stroke
Smash
Slash
Straight Thrust
Recover to Guard Position

Combinations 5
Guard Position
Left Parry a Straight Thrust
Slash
Vertical or Horizontal Butt Stroke
Smash
Slash
Straight Thrust
Recover to Guard Position

Condition 6
Guard Position
Right Parry a Straight Thrust
Straight Thrust
Vertical or Horizontal Butt Stroke
Smash
Slash
Straight Thrust
Recover to Guard Position

Group Strategy

On occasion, you may engage an opponent as a member of a group, or you may engage numerous opponents by yourself or as a member of a group. By combining bayonet fighting movements and simple strategies, you can effectively overcome your opponent or opponents.

Offensive Strategy

Two Against One. If two bayonet fighters engage one opponent, the fighters advance together.

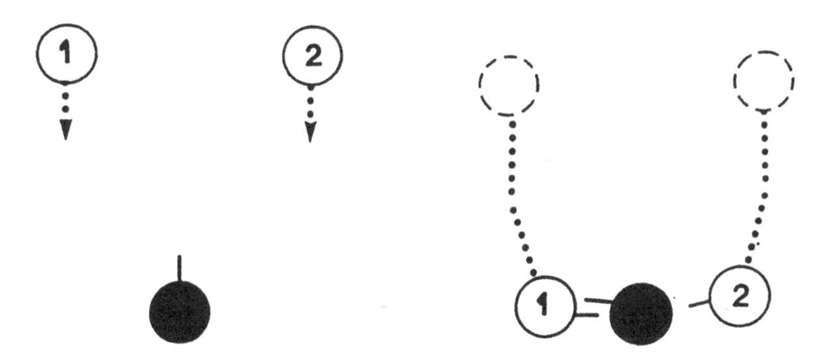

Fighter 1 engages the opponent while fighter 2 swiftly and aggressively attacks the opponent's exposed flank and destroys him.

Three Against Two. If three bayonet fighters engage two opponents, the fighters advance together keeping their opponents to the inside.

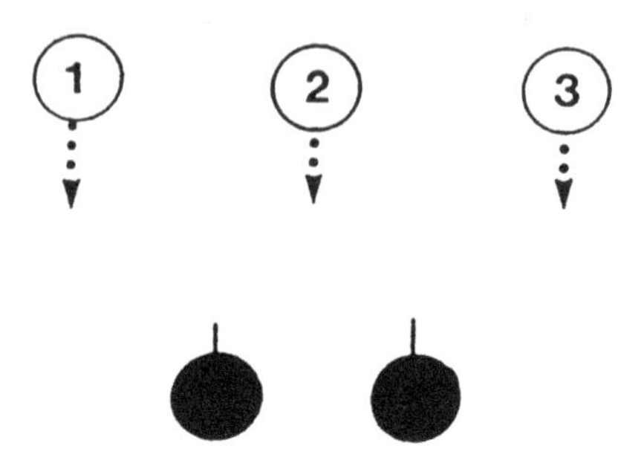

Fighters 1 and 3 engage the opponents. Fighter 2 attacks the exposed flank of the opponent engaged by fighter 1 and destroys him. Fighters 1 and 2 turn and attack the exposed flank of the opponent engaged by fighter 3 and destroy him.

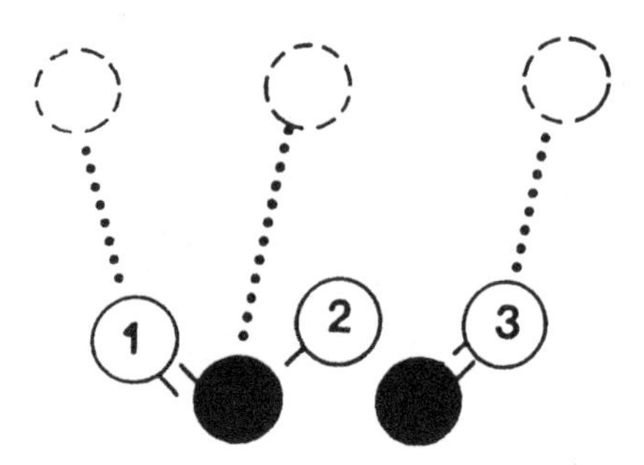

Defensive Strategy

One Against Two. If a fighter is attacked by two opponents, he immediately positions himself at the flank of the nearest opponent and keeps that opponent between himself and the other opponent.

Using the first opponent's body as a shield against the second opponent, the fighter destroys the first opponent quickly before the second opponent can move to assist.

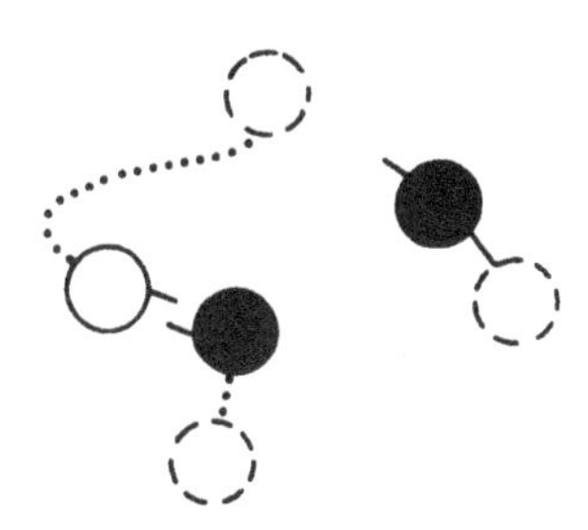

Then, the fighter engages and destroys the second opponent.

Two Against Three. If two fighters are attacked by three opponents, the fighters immediately move to the opponent's outboard flanks.

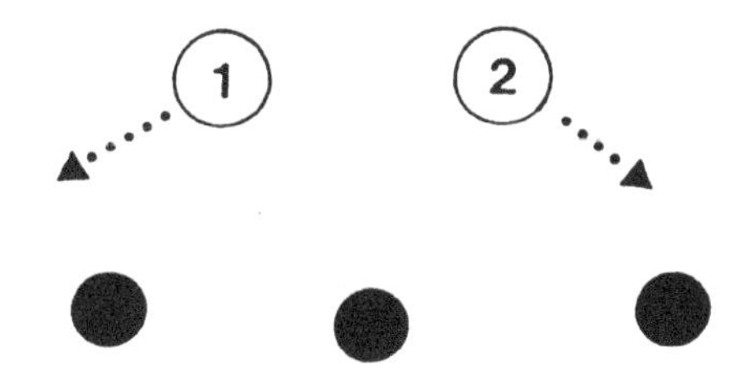

Fighters 1 and 2 quickly attack and destroy their opponents before the third opponent can close in.

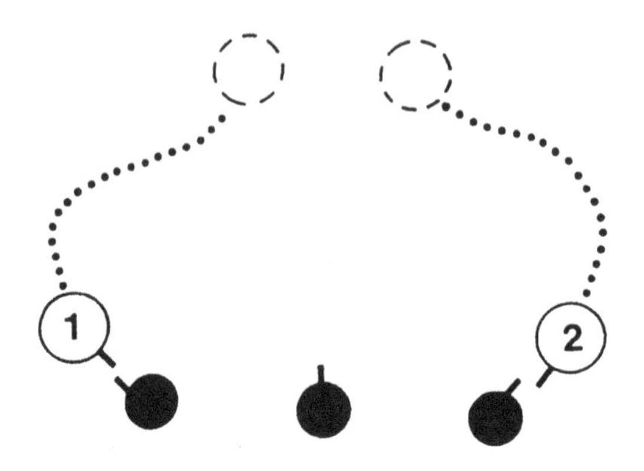

Fighter 1 engages the third opponent while fighter 2 attacks the opponent's exposed flank and destroys him.

LINE VI

Unarmed Defense Against Bayonet Attacks. An opponent armed with a bayonet is a deadly adversary. The first step of unarmed defense against a bayonet attack is to neutralize the bayonet. The most effective way to neutralize the bayonet is to damage the opponent's arms. If confronted with the slash or the straight thrust, you neutralize the opponent's lead arm. If confronted with the smash or horizontal or vertical butt strokes, you neutralize the opponent's rear hand. Once the bayonet is neutralized, you destroy the opponent using the techniques in LINE I, LINE II, and LINE III.

Counter to the Slash

To counter the slash, assume the basic warrior stance. As the opponent attacks with the slash–

- Step in quickly to execute a lead hand parry to the opponent's lead arm.

- Push the opponent's arm away and down to the left of your body.

- Grab the opponent's lead wrist with your lead hand.
- Execute a rear forearm strike to the opponent's elbow to damage and neutralize the arm.

- Apply pressure to the opponent's arm with your forearm to force his head down for the kick to the face.

- Execute a rear leg front kick to the opponent's face.

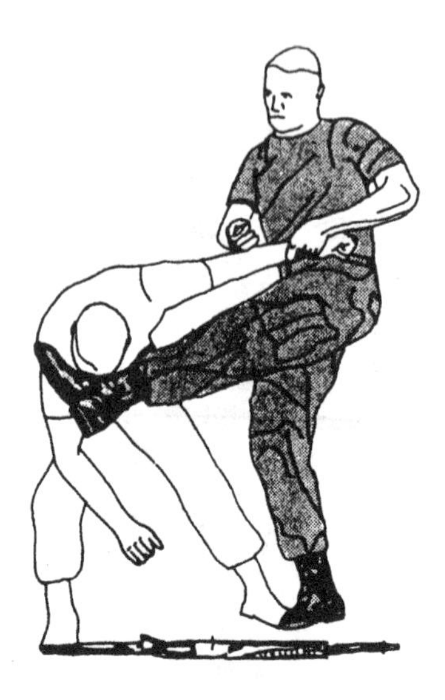

- Grab the opponent behind the neck with your lead hand.
- Grip the opponent's injured arm with your rear hand.

- Rotate your hips and execute a leg sweep to take the opponent to the ground.

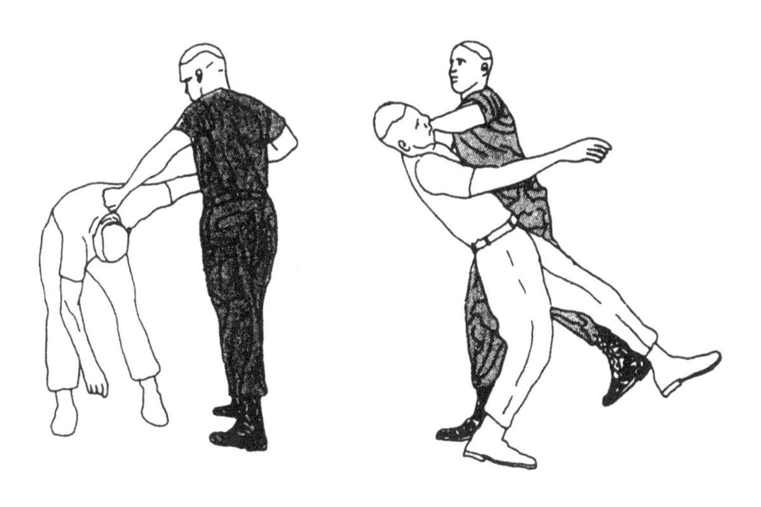

- Execute the heel stomp **swiftly and violently** as a finishing technique.

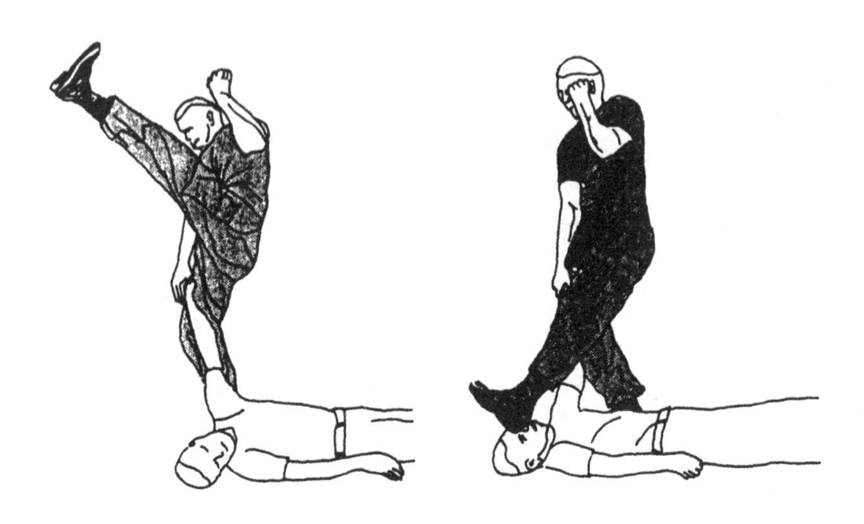

Counter to the Straight Thrust

To counter the straight thrust assume the basic warrior stance. As the opponent attacks with a straight thrust –

- Step to your right quickly.

- Execute a lead hand parry to the opponent's lead arm. This forces the weapon to pass by your left side.

- Grab the opponent's lead wrist with your lead hand.

- Execute a rear forearm strike to the opponent's elbow to damage and neutralize his arm.

- Apply pressure to the opponent's arm with your forearm to force his head down for the kick to the face.

- Execute a rear leg front kick to the opponent's face.

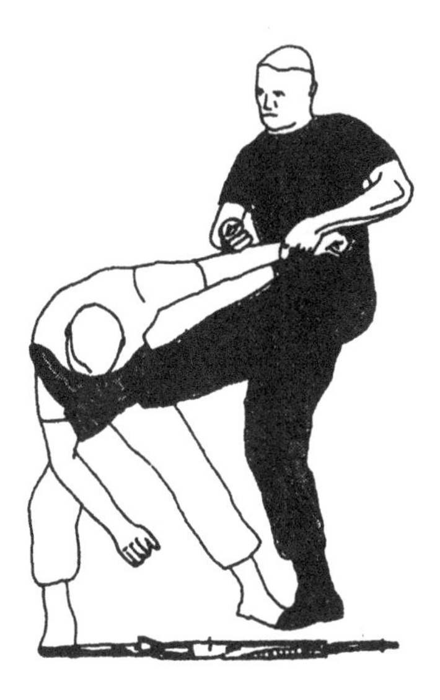

- Grab the opponent behind the neck with your lead hand.
- Grip the opponent's injured arm with your rear hand.

- Rotate your hips and execute a leg sweep to take the opponent to the ground.

- Execute the heel stomp **swiftly and violently** as a finishing technique.

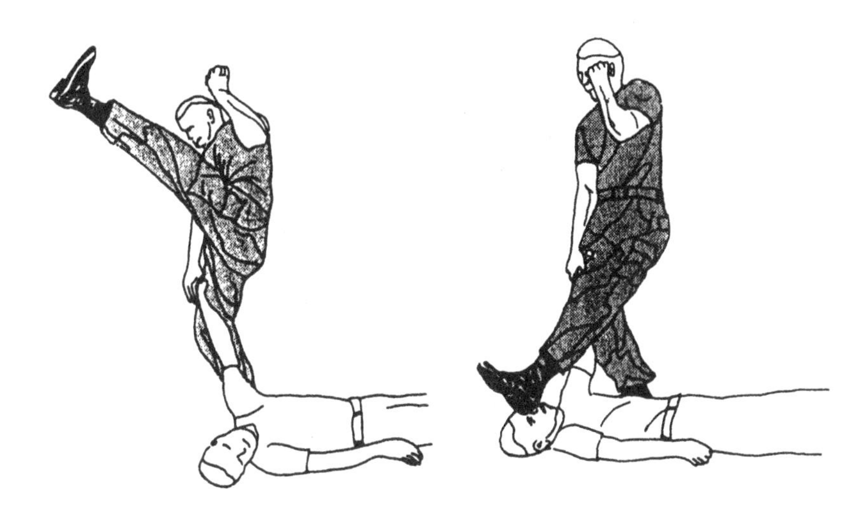

Counter to the Horizontal Butt Stroke

To counter the horizontal butt stroke, assume the basic warrior stance. As the opponent attacks with a horizontal butt stroke—

- Step in and to your left quickly.

- Execute a rear hand parry to the opponent's rear arm.

- Push the opponent's arm away and down to the right of your body.

- Grab the opponent's rear wrist with your rear hand.

- Execute a lead forearm strike to the opponent's elbow to damage and neutralize the arm.

- Apply pressure to the opponent's arm with your lead forearm to force his head down for the kick to the face.

- Execute a rear leg front kick to the opponent's face.

- Grab the opponent behind the neck with your rear hand.
- Grip the opponent's injured arm with your lead hand.

- Rotate your hips and execute a leg sweep to take the opponent to the ground.

- Execute the heel stomp swiftly and violently as a finishing technique.

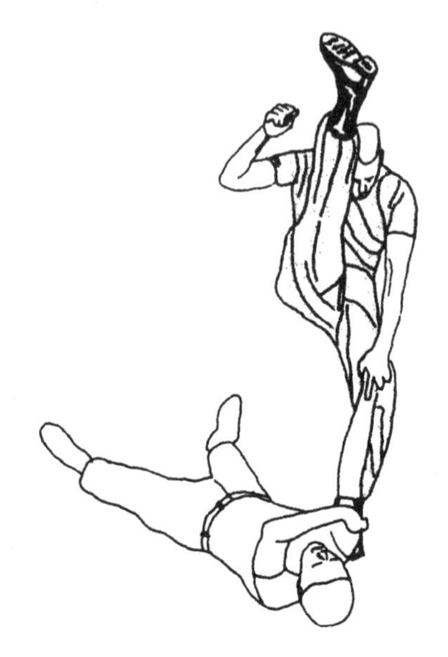

Counter to the Vertical Butt Stroke

To counter the vertical butt stroke, assume the basic warrior stance. As the opponent attacks with a vertical butt stroke—

- Execute a lead hand parry to the opponent's rear arm.

- Push the opponent's arm away and to your right.

- Deliver a palm strike to the inside of the opponent's rear elbow with your rear hand. This creates a gap between the opponent's torso and rear arm.

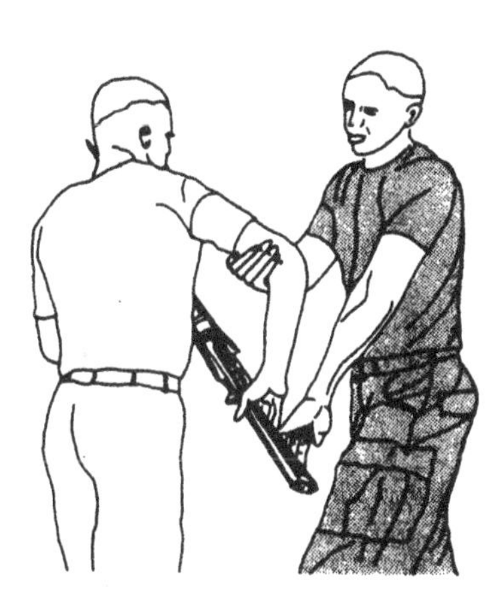

- Move your lead hand under the opponent's rear arm to the back of his neck.

- Release the opponent's arm and rapidly move your rear hand over the opponent's rear shoulder to the back of his neck. This controls the opponent's rear arm.

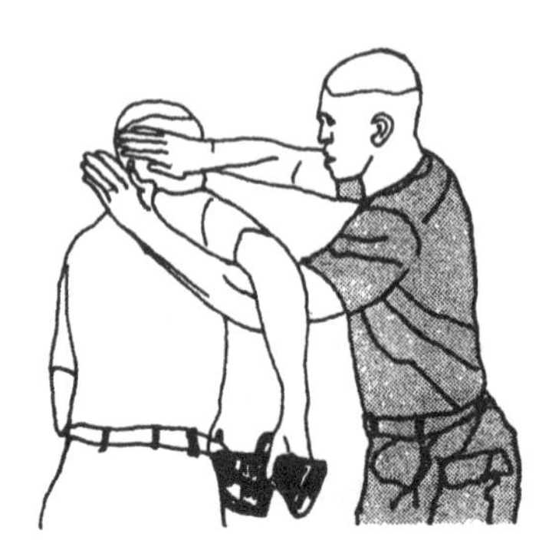

- Apply pressure to the opponent's locked rear arm and neck to force his head down.

- Execute a knee strike to the face.

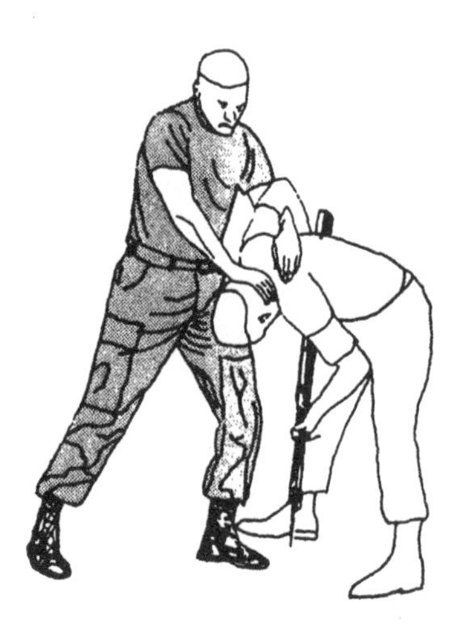

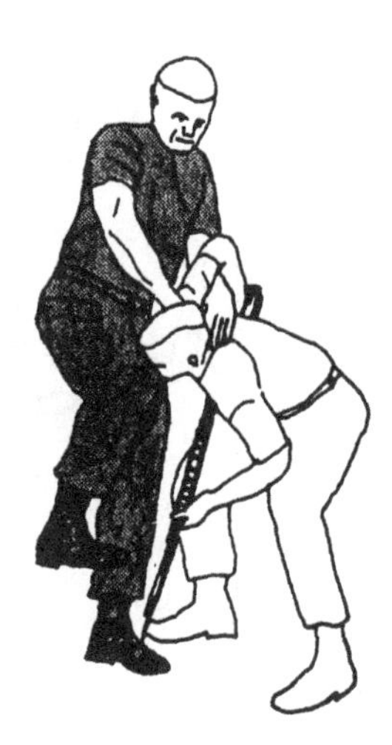

- Grab the opponent behind the neck with your rear hand.
- Grip the opponent's wrist with your lead hand.
- Rotate your hip and execute a leg sweep to take the opponent to the ground.

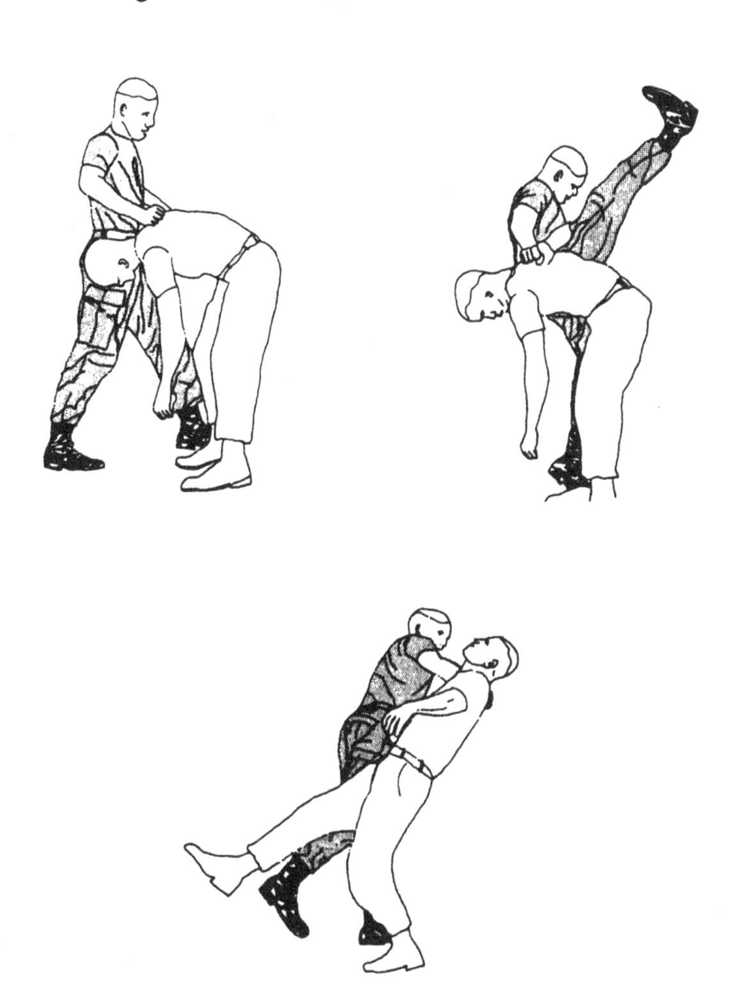

- Execute the heel stomp **swiftly and violently** as a finishing technique.

Counter to the Smash

To counter the smash, assume the basic warrior stance. As the opponent attacks with the smash—

- Step to the left quickly.

- Execute a rear hand parry to the opponent's rear arm.

- Pull the opponent's arm away and down to the right side of your body.

- Grab the opponent's rear wrist with your rear hand.
- Execute a lead forearm strike to the opponent's elbow to damage and neutralize the arm.

- Apply pressure to the opponent's arm with your forearm to force his head down for the kick to the face.

- Execute a rear leg front kick to the opponent's face.

- Grab the opponent behind the neck with your rear hand.
- Grip the opponent's injured arm with your lead hand.
- Rotate your hips and execute a leg sweep to take the opponent to the ground.

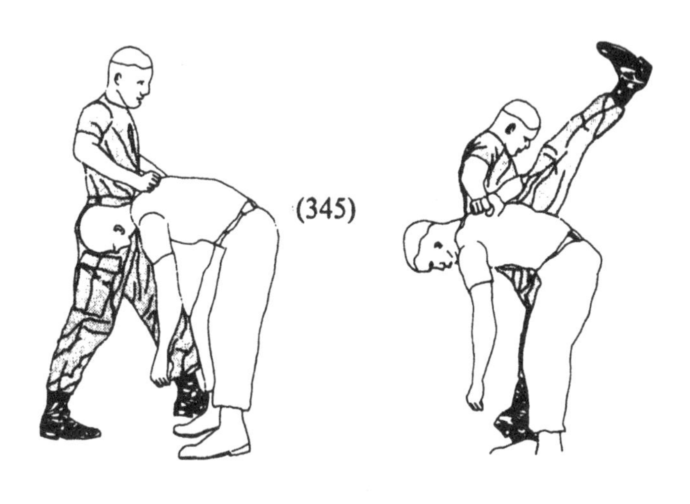
(345)

- Execute the heel stomp swiftly and violently as a finishing technique.

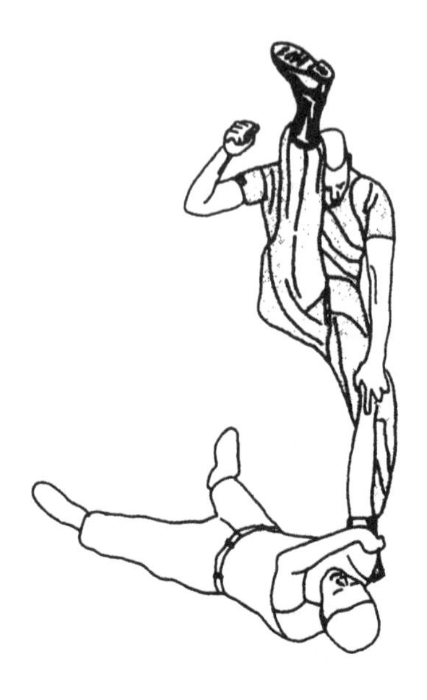

WEAPONS OF OPPORTUNITY

Hand Weapons

A hand weapon is anything that can be held in your hand and used to damage or destroy the opponent. A battlefield can contain a variety of objects that you can pick up and use as a weapon. Your resourcefulness and imagination are your only limitations. This chapter provides examples of how to use a hand weapon if your primary weapon (i.e., firearm, knife, bayonet) is not available.

Entrenching Tool. The entrenching tool (E-tool) is an excellent weapon, especially when sharpened. The E-tool can be used to block and strike the opponent. You can use its sharp edges to slash the opponent's neck/throat area.

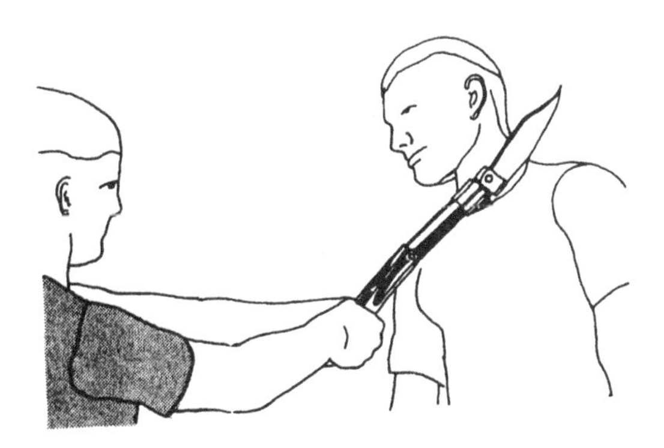

You can use its point to deliver a straight thrust into the opponent's face/throat.

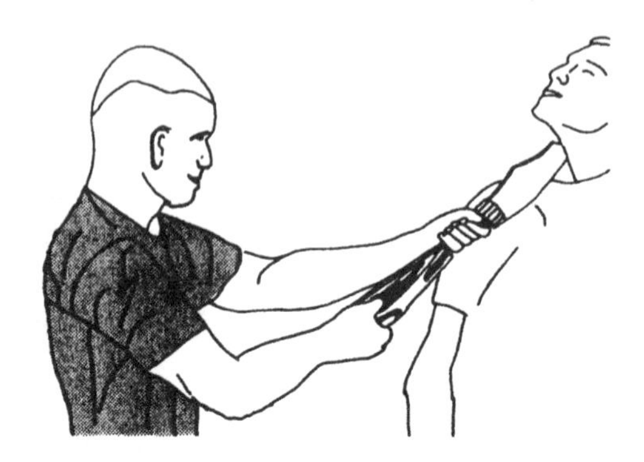

Once the opponent has been forced to the ground, you can use the E-tool to crush the his skull or throat.

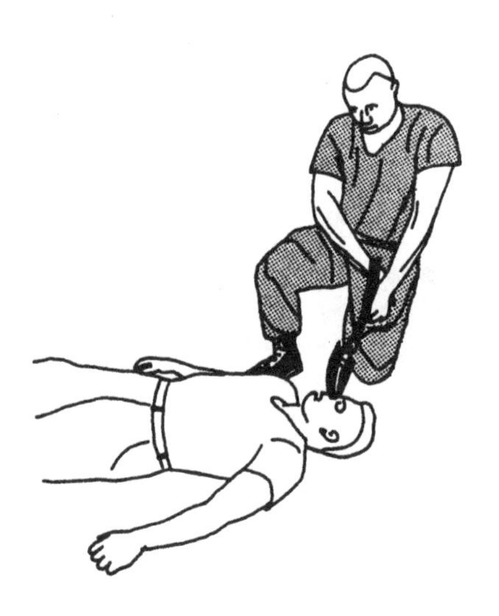

Helmet. You can use a helmet to strike the opponent's unprotected area. The preferred target area is the opponent's head/face.

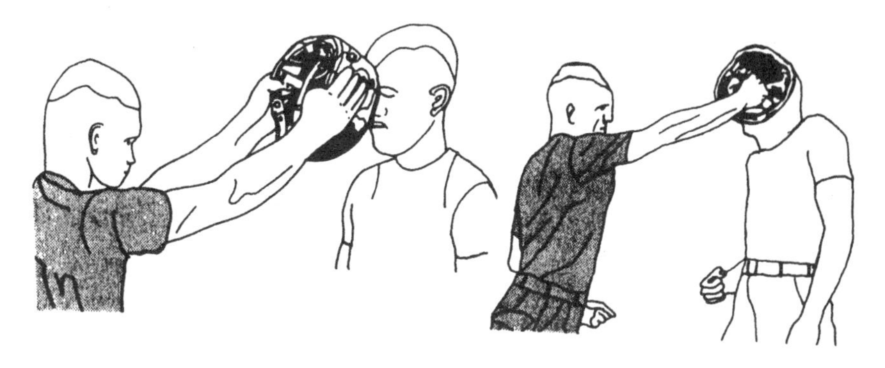

Tent Poles and Pins. You can use tent poles and pins to strike any of the target areas identified in knife fighting. The preferred target area is the throat and groin.

Note

Any sharp, hard object (e.g., broken tree limbs, sticks, iron rods, pipes) can be used in the same manner as tent poles and pins.

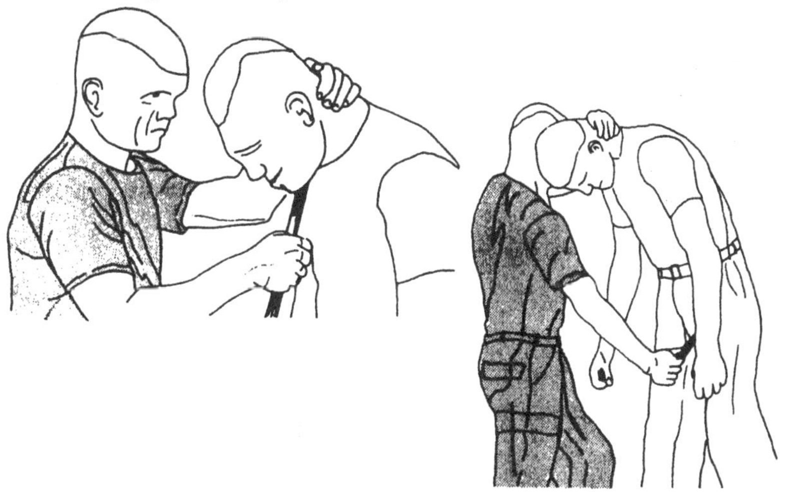

You can use tent poles and pins to block and parry armed or unarmed attacks.

Web Belt. Stretch the web belt between your hands to block attacks. Once the attack is blocked, you can follow up with garroting techniques.

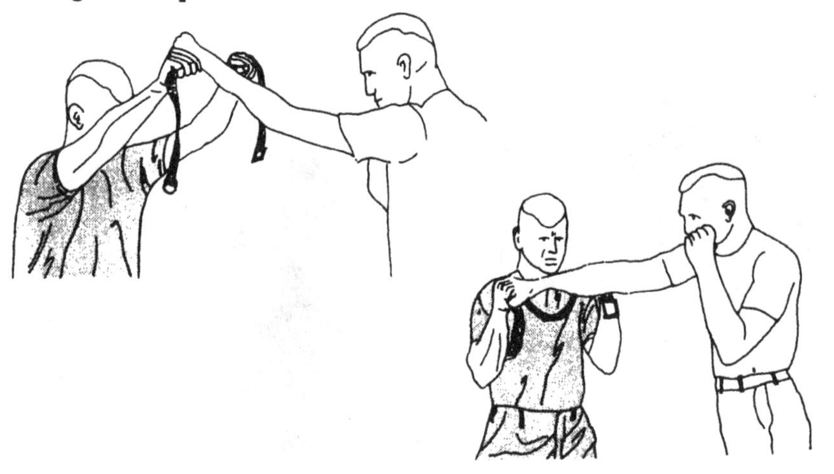

ALICE Pack. You can use the ALICE pack to block or deflect attacks. This allows you time to regain the initiative and destroy the opponent.

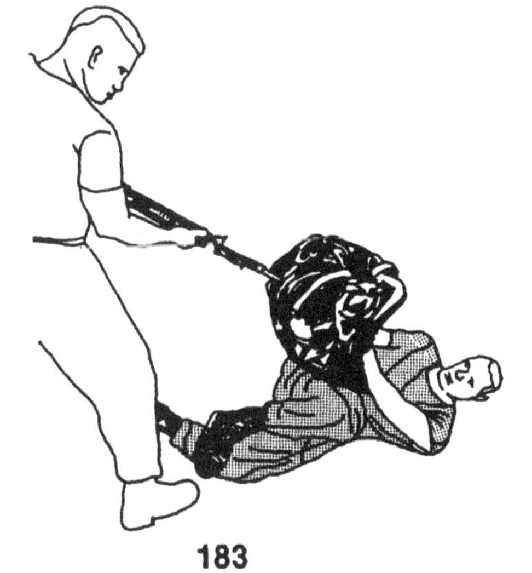

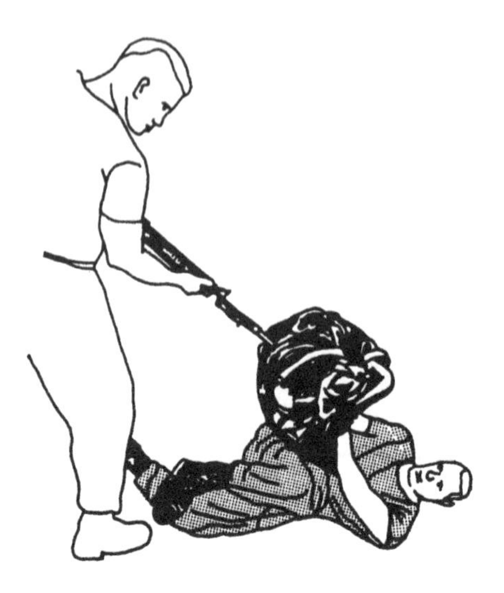

Tent Guide Line, Boot Lace, Communication Wire. Any type of line (e.g., guide line, boot lace, communication wire, barbed wire, etc.) can be used to garrote an opponent. The line is either wrapped around an object (e.g., tent pin or tent pole broken in half) for leverage or wrapped around your hands to establish a firm grip. The line is then stretched between your hands.

If attacking from the rear–

- Stalk the opponent from the rear.

- Keep your body below the opponent's line of sight.

- Crouch in a modified basic warrior stance.
- Extend your arms and drop the line over the opponent's head.

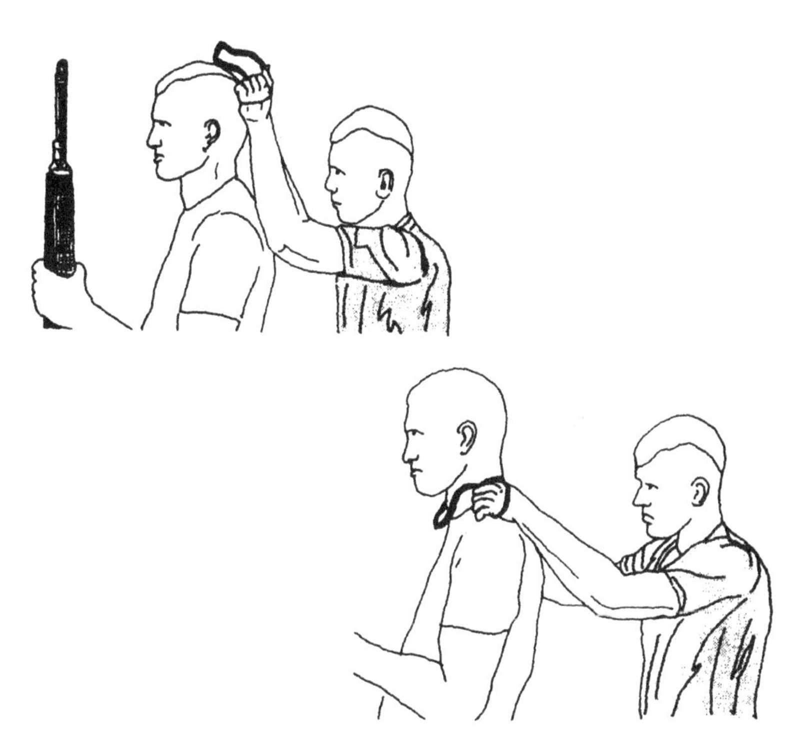

- Step behind the opponent with your rear foot to lock the garrote in place.

- Snap your opponent's arms forward to close his airway and restrict blood flow to the brain.

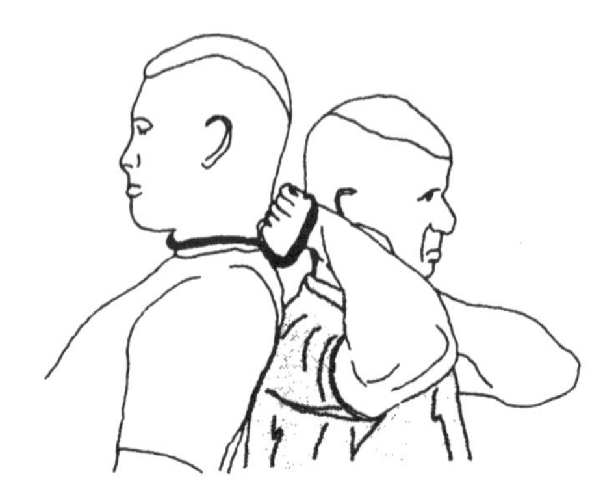

- Keep your body erect and continue to apply pressure until the opponent is eliminated.

Garroting an opponent can also be executed from the front by using the momentum of the attack to gain access to the opponent's rear.

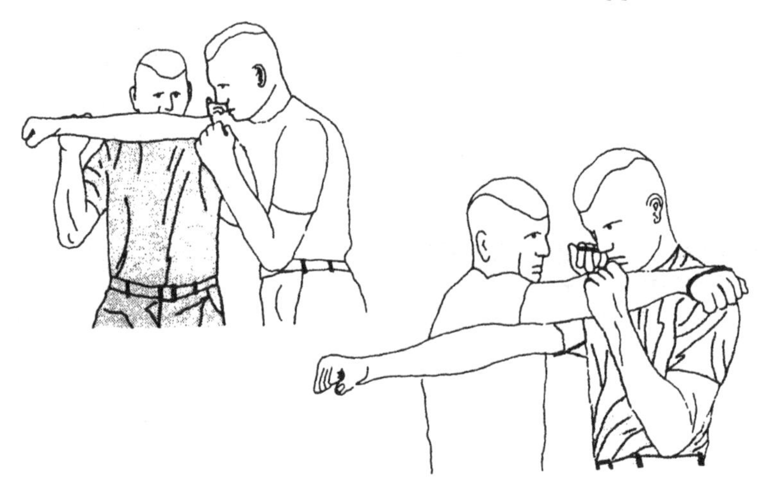

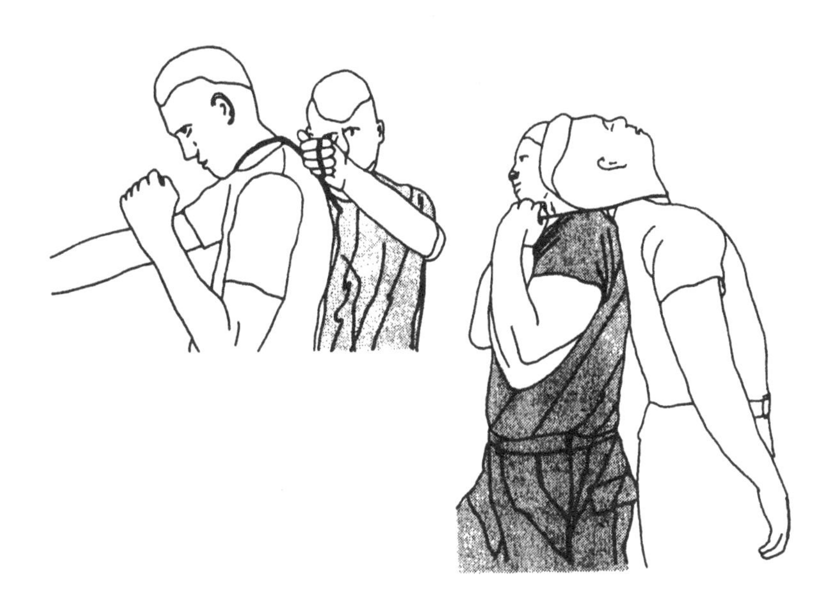

Stationary Weapons

Stationary weapons are objects that you cannot pick up. You can use stationary weapons to immobilize or destroy the opponent by driving him into the object. The terrain, a piece of equipment, a building, a vehicle, etc. can be used by driving the opponent into the object.

The following examples of stationary weapons are only a few of the many possible weapons that can be found on the battlefield. Your imagination and resourcefulness are your only limitations.

Example 1, Tree Trunk. If the opponent attacks with a straight knife thrust—

- Execute an outside block with your lead hand and forcefully push the opponent's arm against the tree trunk.

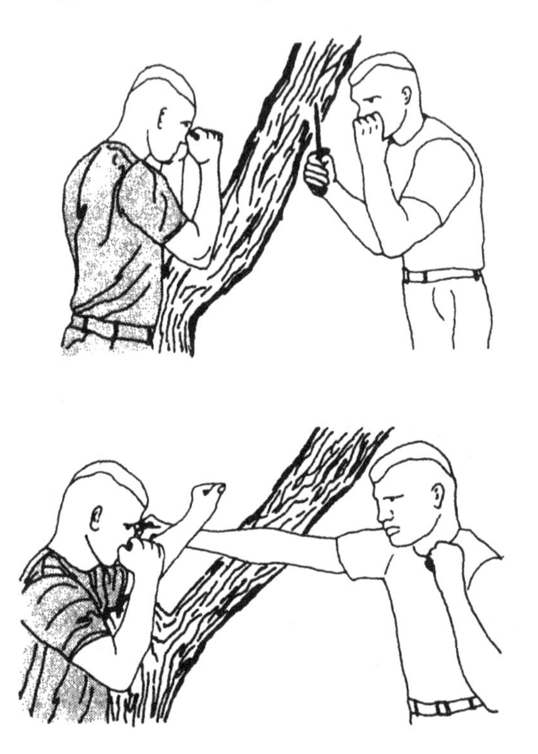

- Apply pressure to the arm with your lead hand and execute a throat grab with the rear hand to crush the trachea.

- Deliver the finishing technique.

Example 2, Tree Limb. If the opponent attacks with a lead hand punch, execute the defense for a lead hand punch described on page 72.

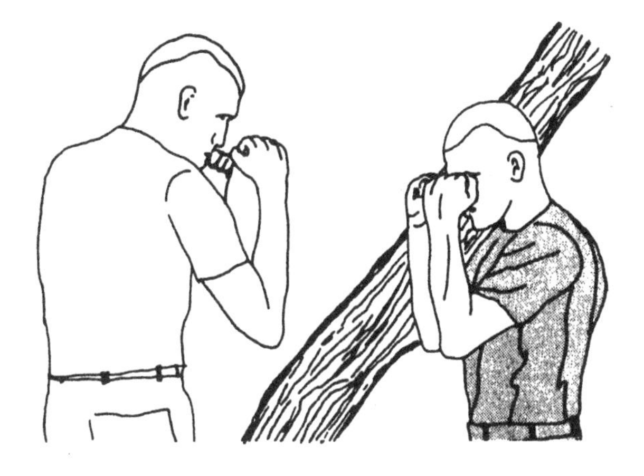

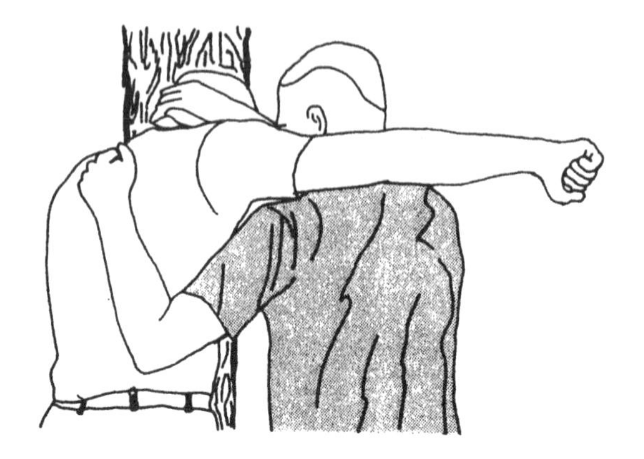

Instead of executing a knee strike to the opponent's face, use the momentum of the attack to impale the opponent's skull on the protruding tree limb.

CPSIA information can be obtained
at www.ICGtesting.com
Printed in the USA
BVHW071647050222
628080BV00006B/700